Lubrication Degradation Mechanisms

Lubrication Degradation
Mechanisms
A Complete Guide

Sanya Mathura, MLE[1]

CRC Press
Taylor & Francis Group
Boca Raton London New York

CRC Press is an imprint of the
Taylor & Francis Group, an **informa** business

---

[1] The MLE (Machinery Lubrication Engineer) certification is a highly coveted certification offered by ICML (International Council for Machinery Lubrication) and Sanya Mathura is the first in her country and the first female to attain it.

First edition published 2021
by CRC Press
6000 Broken Sound Parkway NW, Suite 300, Boca Raton, FL 33487-2742

and by CRC Press
2 Park Square, Milton Park, Abingdon, Oxon, OX14 4RN

Library of Congress Cataloging-in-Publication Data
Names: Mathura, Sanya, author.
Title: Lubrication degradation mechanisms : a complete guide / Sanya Mathura.
Description: First edition. | Boca Raton : CRC Press, 2021. |
Includes bibliographical references and index.
Identifiers: LCCN 2020031639 (print) | LCCN 2020031640 (ebook) |
ISBN 9780367607760 (hardback) | ISBN 9781003102274 (ebook)
Subjects: LCSH: Lubrication and lubricants—Deterioration.
Classification: LCC TJ1077 .M36 2021 (print) | LCC TJ1077 (ebook) |
DDC 621.8/9—dc23
LC record available at https://lccn.loc.gov/2020031639
LC ebook record available at https://lccn.loc.gov/2020031640

ISBN: 978-0-367-60776-0 (hbk)
ISBN: 978-1-003-10227-4 (ebk)

Typeset in Times
by codeMantra

Contents

Preface   ix
Acknowledgements   xi
Author   xiii

1 What Is a Lubricant and What Are Its Functions?   **1**
   1.1   What Is a Lubricant?   1
   1.2   What Are the Functions of a Lubricant?   2
   1.3   Understanding Lubrication Regimes   3

2 Understanding the Types of Degradation Mechanisms   **7**
   2.1   Oxidation   8
   2.2   Thermal Degradation   10
   2.3   Oxidation versus Thermal Degradation   11
   2.4   Microdieseling   12
   2.5   Electrostatic Spark Discharge   13
   2.6   Thermal Degradation versus Microdieseling versus
         Electrostatic Spark Discharge   14
   2.7   Additive Depletion   14
   2.8   Contamination   15

3 Identification of Lubricant Degradation   **17**
   3.1   Basic Tests   17
        3.1.1   Viscosity ASTM D445   18
        3.1.2   Presence of Water/Fuel   19
            3.1.2.1 Water Ingress   19
            3.1.2.2 Fuel Ingress   20
        3.1.3   TAN/TBN   20
            3.1.3.1 TAN   21
            3.1.3.2 TBN   21
        3.1.4   Concentration of Additives   21
        3.1.5   Concentration of Metals   22
        3.1.6   Concentration of Contaminants   23

4 Tests to Determine the Types of Lubricant Degradation   **25**
4.1   Oxidation   25
   4.1.1   Acid Number – ASTM D974   26
   4.1.2   Colour – ASTM D1500   26
   4.1.3   FTIR   26
   4.1.4   MPC (Membrane Patch Calorimetry) – ASTM
       D7843   27
       4.1.4.1   Calorimetric Patch Analyser (CPA)   27
   4.1.5   RULER (Remaining Useful Life Evaluation Routine)   28
   4.1.6   RPVOT (Rotating Pressure Vessel Oxidation Test) –
       ASTM D2272   28
4.2   Thermal Degradation   29
   4.2.1   Viscosity (ASTM D445)   29
   4.2.2   FTIR   30
   4.2.3   Colour (ASTM D1500)   30
4.3   Microdieseling   30
   4.3.1   Visual Inspection of Components   30
   4.3.2   FTIR   31
   4.3.3   QSA (Quantitative Spectrophotometric Analysis)   31
4.4   Electrostatic Spark Discharge (ESD)   32
   4.4.1   FTIR and QSA   32
   4.4.2   RULER   32
   4.4.3   Dissolved Gas Analysis (DGA) – ASTM D3612   33
   4.4.4   Filter Inspection   33
4.5   Additive Depletion   34
   4.5.1   FTIR   34
   4.5.2   Colour   34
   4.5.3   QSA   34
   4.5.4   RULER and RPVOT   35
4.6   Contamination   35
   4.6.1   FTIR   35
   4.6.2   Colour   36
   4.6.3   Presence of Water/Fuel/Coolant   36
4.7   Summary of Degradation Mechanism Tests and Results   37

5 Dealing with Degradation   **39**
5.1   Understanding Your Equipment   39
5.2   Determining the Most Applicable Lab Tests   40
   5.2.1   Understanding the Results   41
5.3   Implementing Measures Based on Lubricant
   Degradation Mechanism   44

5.3.1	Oxidation	44
5.3.2	Thermal Degradation	45
5.3.3	Microdieseling	46
5.3.4	Electrostatic Spark Discharge	46
5.3.5	Additive Depletion	47
5.3.6	Contamination	47

6 Summary   **49**

7 Case Studies   **51**

7.1 Case Study 01: Extending the Lubricant Service Life in
Gas Turbines for Manufacturing Application   51
*Author: Sanya Mathura of Strategic
Reliability Solutions Ltd*

7.2 Case Study 02: Fuel Ingress in Marine Application   53
*Author: Sanya Mathura of Strategic
Reliability Solutions Ltd*

7.3 Case Study 03: Water Ingress in Gas Turbines for
Manufacturing Application   54
*Author: Sanya Mathura of Strategic
Reliability Solutions Ltd*

7.4 Case Study 04: Rapidly Increasing Oxidation in a
Combined Cycle Power Plant   55
*Author: Sanya Mathura of Strategic
Reliability Solutions Ltd*

7.5 Case Study 05: Sporadic Increases in Vibration and
Temperature Levels in an Ammonia Compressor   56
*Author: Sanya Mathura of Strategic
Reliability Solutions Ltd*

7.6 Case Study 06: Lubrication Condition Monitoring Case
Study – Avoiding Critical Asset Damage   58
Author: Andy Gailey of UPTIME Consultant Ltd

7.7 Case Study 07: Allison Transmission on School Buses in Iowa   60
*Author: Michael Holloway MLE, CLS, LLA (I, II), MLT (I, II),
MLA (I, II, III), OMA 1, CRL of 5th Order Industry LLC*

7.8 Case Study 08: Mobile Hydraulic Systems on Mobile Mix
Trucks in Texas   61
*Author: Michael Holloway MLE, CLS, LLA (I, II), MLT (I, II),
MLA (I, II, III), OMA 1, CRL of 5th Order Industry LLC*

7.9   Case Study 09: Industrial Air Compressors in a Glass
      Manufacturing Facility in Texas   62
      Author: Michael Holloway MLE, CLS, LLA (I, II), MLT (I, II),
      MLA (I, II, III), OMA 1, CRL of 5th Order Industry LLC

7.10 Case Study 10: Industrial Hydraulic Systems in an
      Automotive Manufacturing Facility in Indiana   64
      Author: Michael Holloway MLE, CLS, LLA (I, II), MLT (I, II),
      MLA (I, II, III), OMA 1, CRL of 5th Order Industry LLC

References   67
Index   69

Preface

One of the most prevalent challenges that faces the lubrication industry is the degradation of the lubricant, which leads to its failure. There are those who argue that a lubricant does not fail, but it is the machine component that fails. While that argument holds some validity, it should perhaps be rephrased to reflect a more accurate depiction. A lubricant fails when its environment (component, system or process) has been altered to a situation that is no longer conducive to perform its functions.

It is the aim of the author to provide some much needed elucidation on the varying classifications of lubricant degradation. By understanding the processes that the lubricant undergoes during degradation, one can then troubleshoot the reasons behind the actual failure of the lubricant. In essence, this book is aimed at helping readers understand the types of degradation patterns that can occur, typical influencers, methods to detect their presence and ways in which these can be managed, reduced and possibly eliminated from the system.

Preface

Acknowledgements

I would like to thank divine intervention for moving me to pursue my passion for reliability and lubrication engineering. It is with this blessing that this book contributes to the changing reliability engineering landscape. I hope this book serves to inspire any prospective, emerging and devout followers of reliability and lubrication engineering to be the change you wish to see in this world.

Next, I wish to say thank you to the "godfathers" of reliability engineering, Reliability Centre, Inc. Thank you Bob Latino, Mark Latino, Diane Gordon, and the RCI team for welcoming me into the reliability family. This book became a possibility because of your unwavering support, encouragement and our mutual commitment to evolution in reliability engineering.

Undeniably, the tribology industry is stronger together. I extend my sincerest gratitude to fellow industry professionals. Thank you Andy Gailey of UPTIME Consultant Ltd, Michael D. Holloway of 5th Order Industry and of course Cary Forgeron of Bureau Veritas, North America. Your kindness, camaraderie, honesty and willingness to share your wisdom and knowledge with me will forever be etched in my mind.

It is without a doubt that I am humbled to have crossed paths with Greg Livingstone of Fluitec. His continuous work in the Lubrication sector has been an inspiration to me. I am truly grateful for his guidance and willingness to assist with my never-ending questions.

Last but certainly not least, I wish to congratulate my publishers, Taylor & Francis Group for their incredible tenacity and support in making this dream a reality. Your efforts, consistency and commitment to me and my venture throughout this journey will not be forgotten.

Finally, to my twin sister, for being beside me, always.

Author

Sanya Mathura, MLE, is the Founder of Strategic Reliability Solutions Ltd based in Trinidad and operates in the capacity of Managing Director and Senior Consultant. She works with global affiliates in the areas of Reliability and Asset Management to bring these specialty niches to her clients.

Sanya possesses a strong engineering background with a BSc in Electrical and Computer Engineering and MSc in Engineering Asset Management. After completion of her thesis "An investigation into the root causes of lubricant degradation in critical equipment in an Ammonia complex", she was made aware of the plight that reliability faces within and outside of the Caribbean. As such, she decided to ensure that every attempt was made to make reliability the backbone upon which the industry operates by forming Strategic Reliability Solutions Ltd.

She has recently earned her MLE (Machinery Lubrication Engineer) certification from ICML (International Council of Machinery Lubrication) and is the first in her country to secure this certification and notably the first female as well. Sanya has worked in the lubrication industry for the past several years and has used her engineering background to assist various industries with lubrication-related issues both locally in Trinidad and Tobago, regionally and internationally. She has solved lubrication problems and provided training in the Automotive, Industrial, Marine, Construction and Transportation sectors.

What Is a Lubricant and What Are Its Functions?

1

Ideally, we must firstly define a lubricant and understand its functions before delving into its degradation mechanisms. It is of paramount importance that we understand the basics and get these right before we begin to look into the more complex side. This chapter will provide information of the basics. These are the fundamentals upon which the entire lubrication industry is based.

1.1 WHAT IS A LUBRICANT?

A liquid lubricant is defined by Menezes, Reeves and Lovell (2013, 295) as a combination of base oil and additives. It is important to note that there are different types of base oils and additives, and each type has particular properties. As such, the combination of base oil and additives can exist in varying ratios depending on the application of the final lubricant and the lubricant manufacturer.

The classification of base oils is based on the refining method used on the crude oil. According to the Noria (2012b), there are five classifications of base oils as listed below:

- *Group I*: Solvent refined (Mineral)
- *Group II*: Hydrotreated (Mineral)
- *Group III*: Hydrocracked (Mineral)
- *Group IV*: PAO Synthetic Lubricants
- *Group V*: All other base oils not included in Groups I, II, III or IV.

Each base oil group has differing levels of solvency, sulphur, saturates and viscosity index which help to differentiate them. Globally, Group I base oils have been the most common since around 1970 because they were the easiest to refine and more readily available. However, as technology evolved, there has been a trend towards the Group II and higher groups as the value that their chemistry adds to the system has been realized. While these higher groups are costlier than the traditional Group I's, their chemical structures offer more stability and better properties which make them very desirable for use in finished lubricants.

Similarly, each of the additives has its specific function but needs to be formulated to particular ratios when they are being designed for applications. According to Noria (2012a), additives can serve any of the three main purposes as follows:

1. Impart new properties (extreme pressure additives, detergents, metal deactivators and tackiness agents)
2. Suppress undesirable properties (pour point depressants and viscosity index improvers)
3. Enhance properties (antioxidants, corrosion inhibitors, anti-foam agents and demulsifying agents).

Each of the aforementioned purposes acts in tandem with the properties of the selected base oil group to provide the overall characteristics of a finished lubricant. As such, it is critical to fully understand both the properties of the base oil group and additives such that the final characteristics of the finished lubricant can be understood and applied accordingly.

We will cover more details on the functions of the most common additives in Chapter 3 of this book.

1.2 WHAT ARE THE FUNCTIONS OF A LUBRICANT?

Essentially, a lubricant is any material that allows for one surface to slide over the other. For instance, if there was a banana peel on the floor and someone stepped on this, they would slide. In this case, the banana peel acts as a lubricant and allows the human to slide over the floor.

If we speak specifically towards equipment and components, banana peels will not work here. Within the industry, a lubricant must perform five main functions as per Menezes, Reeves and Lovell (2013, 295): reduction of friction, minimizing wear, distribution of heat, removal of contaminants and improvement of efficiency.

Let's think about sharpening the blade of an axe (or any other sharp object). Ideally, we press the blade continuously over the grinding stone in a sliding manner. After some time, the blade starts to get hot, and any uneven surfaces that were present on the blade or the stone begin to get smooth. All of this is done without the presence of a lubricant.

Since the blade is sliding against the stone continuously, it produces friction, which in turn produces heat. At the same time, we have wear occurring between the two surfaces, and any of the filings that came off during this process would remain on the blade or the stone.

If a lubricant was present (such as a grease or an oil), there would be a film between the two surfaces. Now, this film would protect both the blade and the stone, reduce the friction (and by extension the heat being produced), minimize the wear and keep all the filings in suspension, thereby keeping both the axe and the stone clean.

Now, let's take this same example and apply it to components in the facility. Can you imagine what a gearbox with the gears turning against each other (at much higher speeds!) would have to undergo? Therefore, the purpose of a lubricant is to essentially protect the equipment and aid in improving its efficiency during operation. If a lubricant no longer performs any of its five functions, then it has fundamentally failed. We will be exploring these concepts in the upcoming chapters and methods of identification and dealing with these failures.

1.3 UNDERSTANDING LUBRICATION REGIMES

According to Noria (2017), there are four different types of lubrication regimes which can be experienced by surfaces: boundary lubrication, mixed lubrication, hydrodynamic lubrication and elastohydrodynamic lubrication.

Essentially, the best type of lubrication regime is *elastohydrodynamic* as it provides the most ideal environment for both the lubricant and the surface. However, there are instances where the other types of lubrication can exist due to a lack of lubricant or contaminants. Let's explore each type and typical situations in which we can find them.

In *boundary lubrication*, as the name suggests, the surfaces actually touch each other and the oil film does not act as a wedge between them. When we look at surfaces under a microscope, we can see tiny asperities (which we can liken to sharp or jagged edges) which are prevalent along the surface.

Even though a surface may appear shiny and smooth, when we microscopically examine it the actual surface has a lot of asperities. Imagine if two rough surfaces are sliding against each other, like a piece of sand paper against a wall, eventually parts of the wall will be removed due to the asperities of both the sand paper and the wall. Let's translate that on a microscopic level to the surfaces that are to be lubricated.

With an oil film present, this would act as a barrier between the two surfaces and allow them to slide over each other almost seamlessly without damage. However, in *boundary lubrication*, the oil film is not thick enough and the surfaces of the components come into contact with each other. This typically happens during the start-up or shutdown of equipment as the lubricant has not had the opportunity to be fully present to perform its function and can also occur when there are heavy loads at low speeds. According to Noria (2017), as much as 70% of wear occurs during start-up and shutdown of equipment!

Mixed lubrication occurs just after the start-up when the lubricant is now forming a larger film to protect the surfaces. However, it is not fully formed and there are some contact areas where the surfaces will still experience boundary lubrication as well as elastohydrodynamic or hydrodynamic lubrication. During this period, wear occurs due to the areas that are still experiencing boundary lubrication. However, it can be considered a transition phase as the surfaces move from one type of regime into another as the lubricant film gradually increases.

In *hydrodynamic lubrication*, the lubricant provides a film such that both surfaces are adequately separated and their asperities do not come into contact with each other. This is one of the most ideal forms of lubrication as it greatly reduces the wear between the two surfaces and the oil film safeguards that these can easily slide over each other, thus decreasing the friction between them. In this type of lubrication, the oil wedge is maintained in all operating conditions and guarantees that the asperities of both surfaces do not interact with each other.

Conversely, in *elastohydrodynamic lubrication*, the lubricant deforms the contact surface at which it has the highest contact pressure to ensure that the asperities do not touch while maintaining the oil wedge. This type of lubrication usually occurs when there is a rolling motion between two moving surfaces and the contact zone has a low degree of conformity (Noria 2017). Essentially, elastohydrodynamic lubrication occurs when the lubricant allows the contact surface to become elastically deformed while maintaining a healthy lubricant film between the two contact surfaces.

It is important to understand that different lubrication regimes exist. Depending on the regime that is prevalent, we can now recognize whether the lubricant is effectively carrying out its functions or if this can lead to its non-performance and eventual failure.

By understanding the various lubrication regimes, we can now apply this to our knowledge of lubricant degradation mechanisms. For instance, if we know that a surface experiences boundary lubrication, we can now assume that the material may undergo abrasive or fatigue wear. Thus, we can deduce that the lubricant will undergo high-temperature valves and may even become contaminated with the abrasive material. As such, by understanding the lubrication regime, we have more information on the environment in which the lubricant existed, and this will help us in determining the type of lubricant degradation mechanism.

Understanding the Types of Degradation Mechanisms

2

As per definition, a lubricant will fail if it does not effectively perform any of its functions as outlined in the previous chapter. Therefore, we need to examine methods in which the lubricant can fail or degrade to such an extent that it can no longer perform any of its five functions, namely, reducing friction, minimizing wear, distributing heat, removing contaminants and improving efficiency. When a lubricant fails, this can spell disaster for the surfaces which it should be protecting.

A partial function of the lubricant is to be sacrificial. In its sacrificial nature, it can release antioxidants to protect the equipment from oxidation or corrosion. Additionally, its Total Base Number (TBN) reserve can be depleted to reduce the acidity of the oil. Furthermore, the lubricant can have suspended contaminants that may either act as catalysts to reduce or increase its viscosity, thus making the lubricant unsuitable for the application.

Degradation occurs throughout the service life of the lubricant and can even occur due to non-ideal storage and handling practices. However, limits exist whereby the lubricant can be deemed unfit for service. When these limits are reached and the functions of the lubricant can no longer be carried out, it is said to be degraded.

There have been arguments within the industry that there are only three methods of lubricant degradation (as per Barnes (2003, 1536)). However, upon closer examination by Livingstone, Wooton and Thompson (2007, 36), there are actually six methods of degradation with slight variations in the process of degradation. These variations can be critical in determining the next steps forward for any facility personnel as it would identify the areas in which they need to improve or change.

The six methods of lubricant degradation as defined by Livingstone, Wooton and Thompson (2007, 36) are oxidation, thermal breakdown/degradation, microdieseling, additive depletion, electrostatic spark discharge and contamination.

Each of these mechanisms undergoes different environmental conditions and produces varying by-products which set them apart. However, if we class them according to their environmental triggers, we can get four main types. Firstly, a trigger of the presence of oxygen and increase in temperature can lead to oxidation. On the other hand, if there is a rapid increase in temperature (over 200°C), this can result in either of the following mechanisms: thermal breakdown/degradation, microdieseling or electrostatic spark discharge. Furthermore, if we categorize the trigger as the sacrificial nature of additives, then this can result in additive depletion. Lastly, by classifying the environmental trigger as ingress of foreign material, this can result in contamination.

Let's dive a bit deeper into the identification of each of these mechanisms and how they can be told apart.

2.1 OXIDATION

There are commonly two main conditions that must be present for oxidation to occur: oxygen and temperature. Oxidation (as defined by Barnes (2003, 1536)) is the addition of oxygen to the base oil present in the lubricant to form a number of by-products including:

- *Aldehydes*: an organic compound formed by the dehydration of alcohol when the carbonyl group is bonded to at least one hydrogen atom (–CHO) (Chemistry LibreTexts 2020).
- *Ketones*: contain the carbonyl group, but is bonded to two carbon atoms, and hydrogen is not present (–CO–) (Chemistry LibreTexts 2020).
- *Hydroperoxides*: contain the OOH group (–OOH), but do not stem from the carbonyl group. These are typically viewed as unstable (Chemistry LibreTexts 2019).
- *Carboxylic acids*: also contain the carbonyl group, but they have a second oxygen atom bonded to the carbon atom by a single bond (–CO$_2$H) (Chemistry LibreTexts 2020).

Is it important to know what by-products are formed from oxidation? Yes! By understanding oxidation and the by-products, we can now implement tests to determine the presence of these in the lubricant. By extension, if these are present, we can then gauge the rate of oxidation and determine methods to either reduce the rate of its production or eliminate the cause of this reaction.

By diving a bit deeper, we can understand that oxidation is not an instant process. Actually, it occurs in three stages. If we can identify the stage of oxidation, there is a higher probability that we can eliminate the final product from occurring. Thus, the downtime that would occur from machines experiencing oxidation will also be reduced!

Figure 2.1 shows the stages of oxidation as defined by Wooton (2007, 32) and Fitch (2015, 41).

One of the most interesting things about oxidation is that we can actually track its occurrence before developing into the lacquers that cause our equipment to stop working until they are cleaned.

During the initiation process, the lubricant must have a catalyst to produce that initial free radical. If we can clearly identify through oil analysis the catalyst responsible for that free radical production, then we can eliminate the entire oxidation process! In the next chapter, we will cover the tests that can be used to determine the presence of this mechanism and the best methods in which they can be implemented.

When more free radicals are formed during the propagation phase, this typically indicates that a series of chemical reactions are ongoing. When these reactions occur, the temperature of the oil also increases. As such, one of the tell-tale factors for oxidation is a rise in temperature. This rise in temperature may often be seen as a sawtooth graph if the temperature is being monitored constantly. One can notice that there are repeated spikes of increases (when the chemical reactions occur) followed by sudden decreases.

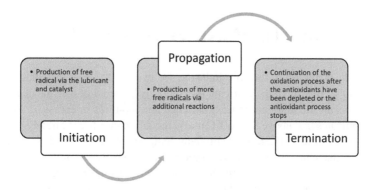

FIGURE 2.1 Stages of Oxidation.

As free radicals are being formed, the lubricant is also doing its best to neutralize these radicals. The lubricant will release its antioxidants to combat these free radicals. As such, the antioxidant levels will start decreasing as this is its sacrificial nature. This is typically seen in turbine oils; the two very useful tests that indicate the presence of antioxidants in the lubricant are the RPVOT and RULER tests. This will be covered in more detail in the next chapter.

Since the process of oxidation has been covered, we need to identify the by-products of oxidation. These will help us in determining the catalysts involved in the oxidation process. The by-products of oxidation are commonly varnish and sludge. However, according to Livingstone, Wooton and Thompson (2007, 36), the properties that the lubricant undergoes due to oxidation can include increase in viscosity, base oil breakdown, additive depletion and loss of antifoaming properties.

As such, when identifying the presence of oxidation, we can look into the following signs:

• Environmental conditions (presence of oxygen and increase in temperature)
• By-products (presence of varnish and sludge) and products that are seen during the process of oxidation (aldehydes, ketones, hydroperoxides and carboxylic acids)
• Changes in properties of the lubricant (increase in viscosity, base oil breakdown, additive depletion and loss of antifoaming properties).

In the next chapters, we will dive into greater detail on the types of laboratory tests that can be performed and methods to handle oxidation depending on the results.

2.2 THERMAL DEGRADATION

As a rule of thumb within the industry, for every 10°C rise in temperature above 60°C, the life of the lubricant is essentially halved. For thermal degradation to occur, the lubricant must experience temperatures in excess of 200°C. This indicates that above the 200°C mark, the lubricant's life is already diminished by 2^{14} times (16,384)! Ideally, any temperature above 60°C requires a lubricant that can withstand these temperatures (typically a synthetic base oil).

Once the temperature of the lubricant exceeds its thermal stability point, then thermal cracking of the lubricant will occur (Barnes 2003, 1536). Just like oxidation, thermal degradation has to undergo a development. As such, once the thermal stability point is exceeded, the hydrocarbon begins decomposition. During this process, small molecules are cleaved off. These molecules can undergo two routes; it either volatizes or condenses. If it volatizes, then there is no deposit formation in the lubricant. However, if it condenses, dehydrogenation commences in the absence of air. With this condensation, coke is usually formed as the final deposit, while other types of deposits are formed along the way before it reaches its final state.

Two critical points to note about thermal degradation are that the lubricant's temperature must exceed its thermal stability point, and if this occurs in the absence of air, by-products will be formed. Due to the nature of the by-products formed, one can establish that thermal degradation has occurred and further investigations can be conducted to determine the source of increase in temperature. This increase in temperature can be either process related or system related.

2.3 OXIDATION VERSUS THERMAL DEGRADATION

The terms oxidation and thermal degradation have been used interchangeably within the industry when describing lubricant degradation. However, these two mechanisms yield different by-products, and their causes must be treated separately to reduce their effects. For instance, one of the main differences is the change of viscosity between the two mechanisms; oxidation yields an increase in viscosity, while thermal degradation produces a decrease in viscosity. Oxidation has an increase in viscosity as the lubricant undergoes polymerization; however, in thermal degradation, the lubricant undergoes shearing which reduces the viscosity.

Figure 2.2 is a brief comparison table of the conditions required and results of oxidation versus thermal degradation.

As seen above, the by-products of both mechanisms vary widely. While oxidation produces by-products that involve molecules reacting with oxygen, thermal degradation by-products are formed from the lack of oxygen. It is important to understand that these two mechanisms are dissimilar, and as such, treatment for reducing their presence will also differ greatly.

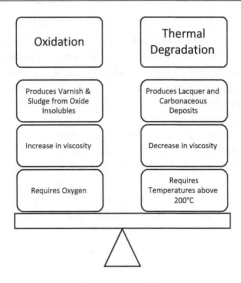

FIGURE 2.2   Oxidation versus Thermal Degradation.

2.4 MICRODIESELING

Microdieseling is also known as compressive heating and is a form of pressure induced thermal degradation. However, the process of microdieseling is quite different from that of thermal degradation, and thus, their by-products are also different.

In this process, entrained air transitions from a low-pressure zone to a high-pressure zone. During this time, localized temperatures in excess of 1,000°C are produced with this transition. The bubble interface becomes carbonized due to the temperatures that it carries when moving through the oil. Lastly, the oil darkens rapidly and produces carbon deposits due to oxidation (Barnes 2003, 1536).

Since this process involves entrained air travelling from the low-pressure zone to the high-pressure zone, there must be a point at which there is an implosion. The type of implosion determines the by-products from this method of degradation. The implosion generally takes place at a low flashpoint with either a low or high implosion pressure.

When there is a high implosion pressure, ignition products of incomplete combustion form which are typically deposits of tars, soot and sludge. On the other hand, if there is a low implosion pressure, then adiabatic compressive thermal heating degradation takes place. This results in the formation of varnish from coke insolubles such as coke, tars and resins. Hence, there is a variation in the products formed depending on the type of implosion pressure that the lubricant experiences.

Thus, while microdieseling is a form of thermal degradation, the final by-products, as well as the process that the lubricant experiences, differ. In microdieseling, the main factor is the transition of entrained air and the formation of incomplete ignition products or coke insolubles (depending on the implosion pressure).

2.5 ELECTROSTATIC SPARK DISCHARGE

Static electricity occurs in daily life when two surfaces are rubbed against each other, and there is a transfer of negatively charged electrons. A similar effect happens with oil molecules. Electrostatic spark discharge occurs at a molecular level with static electricity when dry oil passes through tight clearances (Barnes 2003, 1536). Essentially, static electricity builds up to the point where it produces a spark. This spark can induce temperatures above 10,000°C.

There are three main stages to electrostatic spark discharge (Wooton 2007, 32; Fitch 2015, 41). Firstly, static electricity builds up to produce a spark. At this point, temperatures are in excess of 10,000°C, and the lubricant will begin to degrade significantly. Thereafter, since the spark has started off the reaction, free radicals are then formed. These contribute to the polymerization of the lubricant which accounts for the increase in viscosity. Lastly, the lubricant begins to undergo uncontrolled polymerization. Usually, this leads to production of varnish and sludge which may either remain in solution or become deposited on the surfaces. This uncontrolled polymerization results in the elevation of fluid degradation and increases the presence of insoluble materials.

While the release of free radicals to eventually contribute to polymerization of the lubricant greatly resembles the oxidation process, the initiation stages of electrostatic spark discharge (ESD) greatly differ from that of oxidation. Additionally, ESD occurs with a temperature over 200°C, which allows it to be classified under a form of thermal degradation, even though the lubricant experiences an increase in viscosity compared to the decrease in viscosity seen with thermal degradation. As such, the methods of handling ESD will also contrast those implemented for oxidation. Thus, it is critical to understand and properly identify the type of degradation that is occurring so that it can be reduced or eliminated.

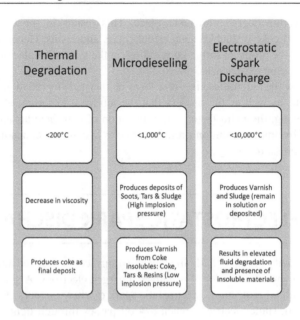

FIGURE 2.3 Comparison of Thermal Degradation, Microdieseling and Electrostatic Spark Discharge.

2.6 THERMAL DEGRADATION VERSUS MICRODIESELING VERSUS ELECTROSTATIC SPARK DISCHARGE

While microdieseling and ESD are two types of thermal degradation, we must understand the differences amongst these mechanisms. The main differences centre around the type of deposits formed and the temperatures that are experienced by each of the mechanisms. Figure 2.3 provides a brief summary that can assist differentiating these three mechanisms.

2.7 ADDITIVE DEPLETION

Earlier, we mentioned the sacrificial nature of lubricants. One of the purposes of additives is to protect the base oil. As such, when degradation occurs, the additives are ordinarily the first to be used depending on the type of degradation.

When deposits are formed, the nature of the deposit is highly dependent on the type of additive being depleted and/or the reaction of the additive with other components in the oil (Barnes 2003, 1536).

Additive deposits can be classified as either organic or inorganic in nature. Understanding this aspect helps us to comprehend how the additive dropout can be affected if it comes into contact with other process elements. Typically, inorganic additive dropouts are additives that have been released from the oil that does not react with anything. One example of this type of additive is Zinc Dialkyl Dithiophosphate (ZDDP) which is used as an anti-wear additive. Conversely, organic deposits usually react to form primary antioxidant species, and these are regularly seen with rust and oxidation additive dropouts.

2.8 CONTAMINATION

Any foreign material that has entered the lubricant and is being used as catalysts can be considered contaminants (Barnes 2003, 1536). These contaminants can be classified into three major groups: metals, water and air. With any of these materials, the process of degradation can be accelerated through oxidation, thermal degradation or microdieseling. While these are large classifications, we must be aware that they can ingress the lubricant in many ways if we are not cognizant of their entry points which are not limited to only outer entry but system processes as well.

One common example is the transferring of lubricants from their original packaging to a smaller pack size and then decanting this into the equipment. We must consider that once the packaging is opened, it is no longer sealed and we have exposed the lubricant to air. When we transfer this lubricant from its original packaging to another pack form (decanter), we can essentially introduce contaminants into the lubricant if the decanter is not a super clean container or was previously used to decant another substance. After we transfer the lubricant from the decanter into the machine, we would have opened the machine (let air in) and then decanted the lubricant.

The simple act of decanting lubricant into a smaller packaging size to be carried to the equipment in the field can introduce contaminants throughout the entire process especially if the environment contains a lot of dust (typically in mining operations). One method of reducing this level of contamination is to have dedicated, sealed decanters or filtration pumps that can be used to filter the lubricant from its packaging directly into the equipment.

As suggested earlier, contamination can also occur within the system during the process. This can transpire when the lubricant comes into contact with an exposed part of the process (such as a leak which can be ammonia or fuel). Additionally, metal can also ingress if there is wear present in the system and the worn metal becomes deposited into the lubricant. Methods of controlling contamination are highly dependent on understanding the types of ingress that can occur within and outside of your system.

Identification of Lubricant Degradation

3

Lubricant degradation can be a facility's worst nightmare as it frequently involves unplanned downtime, rushed logistics, availability of labour and an angry finance department. During these times, budgets usually go out of the window, and when the balance sheets come in at the end of the year, there is always some concern about the numbers after they've been spent. Luckily, there are ways to identify the presence of degradation, monitor its rate of development and possibly eliminate its effects while allowing planned downtime (if necessary) to avoid after-budget meetings.

3.1 BASIC TESTS

When performing routine oil analysis, there are a few basic tests that can help indicate if the lubricant is undergoing degradation. These results are the first indicators, and follow-up tests should be conducted before determining the type of degradation present.

Most labs offer basic oil analysis packages which include the following tests: viscosity, presence of water/fuel, TAN/TBN, concentration of additives, concentration of metals and concentration of contaminants.

The results of these tests cannot be considered in isolation. For instance, if we see a decrease in viscosity, it doesn't automatically mean that thermal degradation is present. However, if there is a decrease in viscosity and an increase in the presence of fuel, we can then determine that the ingress of fuel is responsible for the change in viscosity. As such, we need to investigate the source of the fuel ingress.

Each lab has developed its own internal limits for various tests. These limits have been determined through years of experience and working with

Original Equipment Manufacturers (OEMs). Depending on the application, each test will have a varying limit. Here is a brief overview of the results of each of these tests and what they can indicate about your lubricant.

3.1.1 Viscosity ASTM D445

When evaluating whether the viscosity of the lubricant is still within working limits, one should compare the test value to that of its original viscosity. Viscosities are ranked within ranges and then given a value. Thus, an oil with an ISO viscosity of 46 could have an actual viscosity of 50 cSt or even 42 cSt.

For instance, ISO-rated viscosities observe a ±10% change for their ranges (@40°C) as shown in Table 3.1, which compares three common ISO grades.

As such, if the value obtained through used oil analysis approaches either of these limits, then the oil is no longer regarded as its original weighting and will fail in its function. However, OEMs may have different guidelines regarding viscosity as the engines/components could be built to tolerate viscosities of varying grades. For instance, Caterpillar recommends +20% or −10% of the nominal SAE grade, while Cummins recommends ±1 SAE grade or 4 cSt from new oil viscosity (@100°C). On the other hand, Detroit Diesel advises +40% to −15% of nominal grade (@40°C).

When investigating the suitability of the viscosity of a lubricant after oil analysis (ASTM D445, @40°C), we can consider the following:

- Increase in viscosity (+5%) can indicate that polymerization may be present or solubles may have formed in solution and/or deposited
- Decrease in viscosity (−5%) can indicate that the oil has become sheared, possible contamination with a lighter solution or the presence of thermal degradation.

These are general considerations when investigating the viscosity of a lubricant and should not be used as a standalone result to identify the type of degradation that may have occurred.

TABLE 3.1   Comparison of ISO Grades

ISO GRADE	MIN (CST @40°C)	MAX (CST @40°C)	−5% (CST @40°C)	+5% (CST @40°C)
32	28.8	35.2	30.4	33.6
46	41.4	50.6	43.7	48.3
68	61.2	74.8	64.6	71.4

From Table 3.1, we can clearly see that if the viscosity reaches the ±10% value, then it will no longer be regarded as that ISO grade (@40°C). As such, an indicator of ±5% of the viscosity would be a good range to identify if the lubricant is undergoing any noticeable changes. This is usually seen as the early warning limit. However, if it reaches to the ±10% value, then the lubricant should be changed immediately or as advised by the OEM which is seen as the maximum warning limit.

3.1.2 Presence of Water/Fuel

The presence of either water or fuel can be viewed as both a contaminant and in some cases a catalyst for other reactions. However, both substances have different effects on the lubricant. For instance, water increases the viscosity (through oxidation), while fuel reduces the viscosity of the lubricant (through dilution). Therefore, it is essential when analysing the results that the reasons behind the increase or decrease of viscosity is determined before implementing methods to address these changes.

3.1.2.1 Water Ingress

When testing for the presence of water, labs typically perform two tests: Crackle test and Karl Fischer. In some cases, depending on the type of lubricant (mainly engine oils or transmission oils), labs use the Crackle test as the first indicator of the presence of a considerable amount of water. This involves placing the lubricant on a hot plate. If a crackle is heard, then the level of water is too high and the Karl Fischer test is done to quantify the amount of water present in ppm (parts per million).

In the field, this crackle test can also be done to quickly identify if water is present in the lubricant. The tester can put some of the lubricant in a metal spoon and place a flame under it; they must be very aware on whether flames are allowed in their designated areas. If a Crackle is heard, then there is too much water in the lubricant, and it should be sent for proper oil analysis to determine and trend the quantities of water present. Alternatively, if the user knows that there is too much water in the lubricant, they can begin methods of removing water through a heated filtration process if the equipment cannot be shut down.

When water enters the lubricant, it can remain in any of the following three states: dissolved water, saturated water or excess water (Sedelmeier 2012). *Dissolved water* has a concentration of <500 ppm but remains in the lubricant. *Saturated water* has a concentration that falls within the range from >500 to <1,000 ppm. This allows water to be suspended in oil to form an emulsion.

Excess water can only exist if its concentration is >1,000 ppm. With these levels of water, it can exist as either free water+emulsified oil or free oil+emulsified oil.

Most labs use a condemning limit of 0.3% (3,000 ppm), but 0.1% (1,000 ppm) for turbines. These values can vary greatly depending on application, OEM and lab.

3.1.2.2 Fuel Ingress

Typically, fuel dilution is seen in engine oils, and its presence in high quantities indicates that fuel is entering part of the system that it shouldn't. This can either mean that there is a leak in the actual system or this contamination could have occurred during top-up of the lubricant. One common fuel contamination method with fuel is using the same container to decant both fuel and the lubricant during top-up.

While in some cases, the presence of fuel is not alarming (rather, it forms part of the process), one should always trend these levels to identify whether there is a consistent concentration or if the value fluctuates. This can give an indication of malfunctioning injectors (for engines) or leaks in the actual fuel line.

Fuel dilution can cause the viscosity levels to rapidly decrease, and as a result, wear can occur inside the component as the lubricant can no longer maintain its film. This is considered contamination under the lubricant degradation mechanisms.

Some labs use a warning limit of 2% (20,000 ppm), while others (habitually for diesel engines that are idle for extended periods) may use 3% (30,000 ppm). On the other hand, Cummins uses a limit of 5% (50,000 ppm) before beginning investigations into the causes for the ingress of fuel. Ideally, we should always investigate the source of ingress of fuel as it can decrease the viscosity of the lubricant and possibly raise/lower the flash point of the lubricant.

A field test that can be used to determine the presence of fuel in the lubricant is to place a drop of the lubricant from the system onto a coffee filter and let it "dry" for a couple of minutes. If fuel is present, the viewer will see a rainbow ring to the edges of where the drop of used lubricant has spread out on the filter paper. While this cannot determine the actual concentration of the fuel present in the lubricant, it can help give an indication of the existence of fuel.

3.1.3 TAN/TBN

TAN represents Total Acid Number, while TBN represents Total Base Number. Both are measured in mgKOH/g and help in identifying the acidity levels of the lubricant depending on its type. TAN is used for lubricants such as hydraulic, gear and turbine oils. On the other hand, TBN is used mainly with engine oils. We will explore both tests and their warning limits.

3.1.3.1 TAN

TAN measures the level of acidity in the lubricant. The acidity of a lubricant typically indicates whether there have been catalytic reactions occurring in the lubricant or even oxidation. For instance, during oxidation, free radicals are produced. These free radicals add to the acidity of the lubricant. Thus, an increase in TAN can indicate that oxidation is occurring or has occurred.

An increase in TAN of 0.3 mgKOH/g above the original new oil value indicates that the oil is too acidic. However, depending on the application, this limit can be increased. If an oil becomes too acidic, then this can damage the components by causing additional wear.

3.1.3.2 TBN

On the other hand, TBN indicates the alkalinity of the lubricant. Ironically enough, this value helps us to understand if the oil is becoming too acidic. As we discussed earlier, certain lubricant properties are sacrificial, TBN is one of them. As the oil becomes acidic, the TBN value starts decreasing as it releases oxygen hydroxides into the oil to combat the rising acid levels.

The TBN values are of particular importance especially in engines where acidic blow by-products frequently enter the oil. As such, most lubricant manufacturers have varying levels of base number (BN) for the type of fuel being burned. Ordinarily, Heavy Sulfur Fuel Oil (HSFO) produce more acidic products, and as such, an oil with a high BN (typically around 40 or 50) is used for these applications. However, for regular diesel vehicles (even those with high sulphur levels in excess of 5,000 ppm), a TBN of 10–15 is the usual range. *A decrease of 50% of the new oil value* is used to determine whether the oil should continue in service or not.

3.1.4 Concentration of Additives

Additives are the defining characteristics of a finished lubricant. Every application has a particular set of additives that will be used. For instance, in an engine oil, manufacturers pay close attention to anti-wear, detergents, viscosity index improvers, and demulsifying agents. These additive packages typically account for 30% of the finished lubricant. On the other hand, for turbine oils, additives only make up 1% of the finished lubricant.

TABLE 3.2   Additives and Their Purposes

ADDITIVE	PURPOSE
Molybdenum	Extreme pressure additive in speciality oils and greases. Can be a corrosion inhibitor in some coolants
Magnesium	Detergent, dispersant, alkalinity increaser
Sodium	Corrosion inhibitor in oils and coolants
Boron	Detergent, dispersant, antioxidant in oils and coolants
Barium	Corrosion and rust inhibitors, detergent, antismoke additive in fuels
Phosphorus	Anti-wear, combustion chamber deposit reducer, corrosion inhibitor in coolants
Potassium	Corrosion inhibitor, trace element in fuels, can be a mineral in salt in sea water
Calcium	Detergent, dispersant, alkalinity increaser
Zinc	Anti-wear, antioxidant, corrosion inhibitor
Antimony	Anti-wear, antioxidant

Source: Adapted from Bureau Veritas (2008).

With each application, the concentration of additives will differ. Hence, it is important to have a baseline (or new oil) reference with which to compare the used oil analysis results. It is essential when evaluating whether these additive packages have decreased that we trend the other characteristics of the lubricant as well.

Table 3.2 shows the purposes of various additives adapted from United States. Bureau Veritas (2008).

One example of trending both metals and additive packages can include noticing a steady decrease in zinc/phosphorus (used in the additive ZDDP responsible for anti-wear), while the values of the metals start increasing (such as copper or iron). As the value of the metals may start increasing, we may also notice that the value of the viscosity could be increasing due to the rising concentration of the metals or due to thermal degradation.

We therefore need to fully evaluate the reasons behind the additive depletion before changing out the oil and hoping that this solves the issue.

3.1.5   Concentration of Metals

In the same way that additives have various concentrations depending on their application, metals have this similar stance, whereby depending on the component, the warning limits vary. If we compare the metal warning limits of an engine to that of a transmission or gearbox, we would find that they differ greatly! These warning limits also vary by manufacturer since different components will have various metallurgies.

The warning limit of iron in a Cummins ISX engine is 150 ppm; however, the warning limit for iron in a Caterpillar engine is 35 ppm. On the other hand, the typical warning limit for iron in a gearbox or transmission is 300 ppm. Conversely, the warning limit of iron in a differential is actually 1,000 ppm. Hence, it is critical when evaluating oil reports that we understand the metals that make up the component and their applications.

One field test that can be used to determine the presence of metals in the used lubricant would be the use of a magnet. When performing an oil drain of a sump, we can place a magnet on the underside of the container being used to collect the used oil. Afterwards, we can slowly decant the used oil into another container with the magnet remaining on the underside of the initial container. If metals are present, we may see the filings stuck to one area where the magnet has been placed. This is a great indicator to determine if there are metal filings present; however, a lab would be required to determine their concentrations and actual composition.

3.1.6 Concentration of Contaminants

One of the main contaminants for lubricants is silicon. While contaminants can also include fuel and water (these were covered earlier), there are instances where the ingress of any foreign material is considered a contaminant. In certain instances, this can be grease which may have come into contact with the lubricant either through dispensing or during the actual process.

Typically, silicon is indicative of dirt particles that have entered the oil, and in some cases, it can represent elements of a sealant that was used. For marine engines especially, the concentrations of silicon and aluminium must be monitored as these elements can produce catalytic fines which can degrade the oil faster. The warning limits for silicon is dependent on the application, and it is always important to note the environment in which the equipment is located.

Tests to Determine the Types of Lubricant Degradation

4

To effectively identify the type of lubricant degradation, a series of tests need to be done. These tests cannot be done in isolation, and their results should be trended to establish the trends before determining the type of degradation. As mentioned by Livingstone, Wooton and Thompson (2007, 36), there are a couple of basic tests to identify the presence of degradation and then secondary tests to identify the type of degradation. Here is a brief summary of the types of tests indicative of the type of lubricant degradation.

4.1 OXIDATION

Earlier, we mentioned the by-products of oxidation and the process through which the lubricant would have to undergo during this stage of degradation. As such, there are specific signs that we can look for when identifying if this mechanism is present. Here is a brief overview of the typical tests used for oxidation which include Acid Number (ASTM D974), colour (ASTM D1500), FTIR – Fourier Transform Infrared, MPC (ASTM D7843) – Membrane Patch Calorimetry, RULER – Remaining Useful Life Evaluation Routine and RPVOT (ASTM D2272) – Rotating Pressure Vessel Oxidation Test.

Even though viscosity is not mentioned here, it should also be regarded when evaluating the presence of oxidation. However, viscosity is not a defining test for oxidation, and its increase or decrease can be related to several degradation mechanisms. However, for oxidation, it is typical that the viscosity of the lubricant will increase.

These are some key characteristics that one can look for when identifying the presence of oxidation through the aforementioned tests.

4.1.1 Acid Number – ASTM D974

An increase in the acid number usually indicates that oxidation may be occurring. During the oxidation process, free radicals are formed and the lubricant tries to neutralize these. Consequentially, when the additive package depletes to a state where it can no longer neutralize the acidity, the acidity levels rise as indicated by the acid number. As mentioned earlier, a rise of 0.3 mgKOH/g can be used as an initial warning limit.

4.1.2 Colour – ASTM D1500

With the colour test, the actual colour of the oil isn't measured; instead, the luminance transmittance of light is measured. It is based on a scale of 0.5–8.0 and has 16 intervals.

While the Colour test is not definitive in declaring the presence of oxidation, it is helpful in understanding if oxidation is present (as well as other degradation mechanisms). When oxidation occurs, the oil changes colour and gets darker. One method of tracking if oxidation is occurring is to trend the changes in the colour of the oil such as whether it continues to darken or stays at one level.

4.1.3 FTIR

A lot of people underestimate the power of FTIR (Fourier Transform Infrared). This test actually measures the by-products formed during oxidation (Troyer 2004). These are categorized as either organic or metallo-organic materials which are produced as a result of the oxidation of hydrocarbon molecules.

If an FTIR is done on the insolubles or material that has been deposited in the lubricant, then the wave numbers of various parameters can determine the presence of oxygen or phenol inhibitors (Scott 2013). By determining the presence of these parameters, one can definitely conclude that oxidation has occurred or is occurring.

4.1.4 MPC (Membrane Patch Calorimetry) – ASTM D7843

The MPC test involves measuring the amount of insolubles in a fluid (dissolved or otherwise) through the use of a patch. The patch is rated based on its colour which translates to the amount of insolubles that have been "filtered". There are four general ratings which most labs abide when conducting this test: Good <15, Monitor 15–25, Abnormal 25–35 and Critical >35.

One of the key aspects to remember is that the results of the MPC test have to be trended to understand whether varnish (insolubles) are actually present. There are some cases where these insolubles are present in the system but confined to particular areas. As such, when the oil is sampled from another region, the presence of these insolubles isn't noted. Additionally, some of these may actually remain in solution and not plate out on the surfaces. As such, the MPC test may not quantify its presence. It is therefore very important that this is not used as a standalone test for confirming the presence of oxidation.

Generally, a steady increase in MPC values indicates that oxidation is occurring. This test is particularly useful for rotating equipment. While it may be a bit more pricy than the average lab test, the information that we gain from it can assist us in making that decision to plan maintenance rather than having the equipment shut down and result in unplanned maintenance.

4.1.4.1 Calorimetric Patch Analyser (CPA)

The Calorimetric Patch Analyser (CPA) method was developed after identification of some areas of improvement in the MPC test by Dr Akira Sasaki, Dr Kenji Matsumoto, Dr Tomomi Honda and Mr Greg Livingstone in 2017. The authors presented a paper at the STLE's 73rd Annual Meeting and Exhibition in Minneapolis on May 22, 2018 entitled "A New Calorimetric Method to Detect Varnish Precursors". This paper discusses the method by which MPC patches are analysed and notes that the patches all undergo a heating process during which some of the insoluble materials become soluble. As such, they have identified that there may be a level of inaccuracy in the actual MPC test.

The authors have proposed a new method that measures the colour of contaminants by RGB colour space. It measures colour through the use of reflecting light (for the surface of the patch) and transmitting light for the contaminants captured inside the patch. The final measurement gives an indication of the type of precursor to varnish through comparison with a colour scale. These colours can represent contaminants and precursors such as mild solid particles, severe solid particles, multiple pollution particles, precursors, varnish, sludge or oxidation.

The CPA method is very useful for detecting the stage of oxidation that is occurring in the lubricant through the identification of the types of precursors. Since the stage of oxidation can be detected, we can therefore implement measures to eliminate the causes of oxidation.

4.1.5 RULER (Remaining Useful Life Evaluation Routine)

The purpose of the RULER test is to establish the concentrations of antioxidants in the used oil. Earlier, we spoke about the sacrificial nature of lubricants, in particular the antioxidants. This test can trend the depletion orate of these antioxidants (phenolic and aromatic amines) and compare it against the new oil levels. This comparison can provide us with an estimated remaining useful life of the product, should we have to plan for maintenance.

Ideally, the results of this test should not be examined in isolation. For instance, if the RULER test results indicate that the remaining useful life is >25%, but the RPVOT results indicate that the time remaining is 900 minutes, while the MPC tests come back with a patch rating of >15, then we don't need to change the oil right away. However, if after a second test the value of the RULER drops again significantly, with the values in the other tests fluctuating, then we have a cause for concern.

Usually, with some turbine oils, they undergo a rapid rate of antioxidant depletion on start-up, and then these values taper out to RULER values above 25%. As such, it is critical to trend these results to really understand what is occurring in the oil.

4.1.6 RPVOT (Rotating Pressure Vessel Oxidation Test) – ASTM D2272

As mentioned previously, the RPVOT and RULER tests almost go hand in hand when evaluating the occurrence of oxidation in oils. While the RULER test gives its result in percentage of life remaining, the RPVOT gives a result in minutes. The time recorded is the time taken for the oil to reach a specified pressure drop, indicating that oxidation has occurred. These minutes are representative of the oil's resistance to oxidation and can be compared against new oil values to get an estimate of the rate of oxidation.

While we mentioned that the RULER was a tad expensive, the RPVOT goes up a couple of notches and is one of the most expensive oil analysis tests!

Ideally, the frequency of performing the RPVOT is typically limited to once or twice yearly but can vary depending on what is being trended. Similarly to the RULER test, the warning limit is below 25% of the new oil value. The RPVOT has to be trended over a period of time and decisions about its results should not be done in isolation.

We have developed a brief summary of all the tests used for oxidation and their respective warning limits.

Viscosity: An increase by 5% should be monitored.

Acid number: An increase of 0.3 mgKOH/g is an initial warning limit and should be investigated.

Colour: Should one notice rapid changes in the scale associated with colour, this is concerning.

FTIR: The presence of insolubles can indicate that these will eventually deposit onto the surfaces.

MPC: Values that are above 25–35 are considered abnormal, while those above 35 are critical.

RULER: If the value is >25% new oil, there is cause for concern.

RPVOT: If the value is >25% new oil, this should be investigated.

4.2 THERMAL DEGRADATION

Essentially, there are three main tests to determine if thermal degradation has occurred or not. These include viscosity (ASTM D445), FTIR and Colour (ASTM D1500). While we have covered these three tests in oxidation, we will be on the lookout for slightly different results as indicated below.

4.2.1 Viscosity (ASTM D445)

With thermal degradation, the lubricant often experiences thermal cracking. This leads to the shearing of the lubricant which in turn decreases the viscosity. As mentioned earlier, once the viscosity value starts approaching another range (where it will no longer be classified as its original grade), then we start monitoring closely.

In the instance of thermal degradation, we will be on the lookout for a decrease in viscosity of −5%. However, we also need to examine if there are other reasons for this decrease (such as fuel ingress or another contaminant).

4.2.2 FTIR

In thermal degradation, we typically find coke as a final deposit (due to dehydrogenation). However, other deposits can occur between the onset of thermal degradation and its grand finale. Ideally, we will be on the lookout for lacquer and other carbonaceous deposits, which can be identified through FTIR.

4.2.3 Colour (ASTM D1500)

As indicated earlier, the test for colour has 16 ratings. Habitually, due to the lubricant undergoing high temperatures (in excess of 200°C), the lubricant darkens almost rapidly. Thus, a tell-tale sign would be the rapid darkening of the oil. While all of these tests are included in that for oxidation, two of the key differentiators would be the decrease in viscosity and the presence of carbonaceous deposits.

Here is a brief summary of the tests used for thermal degradation and associated warning limits:

Viscosity: Ideally, a decrease of 5% from the new oil value is concerning.
Colour: If rapid changes occur on the colour scale, this could be indicative of thermal degradation.
FTIR: Since this type of degradation produces carbonaceous deposits, we should be on the lookout for deposits of this nature.

4.3 MICRODIESELING

With microdieseling, the lubricant is subjected to temperatures in excess of 1,000°C and it is a form of thermal degradation. As such, there are a few tests that can be used to verify if the lubricant has experienced this type of degradation. It is also very noteworthy that with this type of degradation, it involves entrained air which can lead to cavitation. Thus, some of the tests that can be used to determine if this is present can include the following.

4.3.1 Visual Inspection of Components

While this is not a test that will always be carried out in a lab, it is a very effective one for the field. With microdieseling, the entrained air undergoes either a high or low implosion pressure. As such, this pressure may occur on the surfaces of the equipment. Therefore, it will be easy to spot if any cavitation is occurring on the components. The presence of cavitation indicates that there is entrained air in the system, and this could be contributing towards microdieseling.

4.3.2 FTIR

As you may have realized, FTIR is one of the most useful and versatile tests. This is mainly since the test allows the presence of different materials to be identified. For instance, with microdieseling, some of the deposits can include soot, tars, sludge (low flashpoint, high implosion) or carbon insolubles such as coke, tars or resins (low flashpoint, low implosion).

Therefore, the presence of any of these deposits, coupled with the visual inspection of the components, can help in identification of the presence of microdieseling.

4.3.3 QSA (Quantitative Spectrophotometric Analysis)

QSA uses a direct correlation between the colour and the intensity of the insoluble products to determine the amount of lubricant degradation that has occurred. When a QSA is performed, it isolates and measures the specific lubricant degradation by-products. Thus, by the identification of by-products, one can determine the type of degradation that has occurred.

For microdieseling, some of the deposits can include soot, tars, sludge (low flashpoint, high implosion) or carbon insolubles such as coke, tars or resins (low flashpoint, low implosion) as per Livingstone and Thompson (2005, 16). Here is a brief summary of the tests for microdieseling:

Visual inspection of components: This is done to identify if cavitation present which is usually caused by the presence of trapped air bubbles.

FTIR and QSA: Both tests can be used to identify the presence of the following by-products: soot, tars, sludge (low flashpoint, high implosion) or carbon insolubles such as coke, tars or resins (low flashpoint, low implosion).

4.4 ELECTROSTATIC SPARK DISCHARGE (ESD)

One of the main challenges with hydraulic and turbine oil users is the passage of oil through very tight clearances. This passage almost always leads to a build-up of static electricity, and when the oil passes through these clearances, it can spark due to this build-up. Typically, this involves temperatures of up to 10,000°C or even higher!

During this process as described in the previous chapter, the lubricant undergoes rapid polymerization, and varnish, sludge or other insoluble materials are produced. One of the best methods of reducing this type of degradation is to decrease the build-up of static electricity. This can be done through reducing the flow rates through the filters, having the filter materials made of non-conducting materials, increasing the residence time of the lubricant to allow time for the static to discharge and implementing a grounding mechanism on the inside surfaces.

One of the common identification sounds for ESD is the sparks that operators hear from the outside of the machines. This is typically a ticking or clicking sound very close to the filters. Many operators hear this on a daily basis but don't pay heed to this noise. This clicking/ticking is the lubricant telling you, "Hey we've got some ESD happening here!" There are five main tests for ESD: FTIR, QSA, RULER, DGA and filter inspection.

4.4.1 FTIR and QSA

These tests, as we have been seeing, can help us identify the presence of some of the by-products typically seen in ESD. With ESD, the by-products formed are varnish, sludge or insoluble materials. When we couple the results of these tests to that of the filter inspection, DGA and RULER, it helps us to definitively identify the presence of ESD.

4.4.2 RULER

As indicated earlier, the RULER test results tell us about the presence of the antioxidants in the lubricant. If these values are being depleted rapidly, then this could be as a result of the free radicals that are produced during ESD. The lubricant will try to neutralize these radicals to reduce the polymerization effect. Thus, with a decrease in these antioxidants, we need to be wary of ESD.

4.4.3 Dissolved Gas Analysis (DGA) – ASTM D3612

According to Livingstone, Wooton and Thompson (2007, 36), lubricants that undergo degradation through hydrocracking of the molecules release particular gases. Depending on the gas (and its concentration) present, one can determine the type of degradation that has occurred.

It was found that acetylene is typically produced when lubricants are exposed to temperatures above 700°C (Lewand 2003, 62). While ESD occurs with a spark in excess of 10,000°C, this temperature is momentary. Additionally, ethylene and methane were found to be present in high concentrations in lubricants that experienced overheating.

Thus, the three gases to pay attention to when performing DGA would be acetylene, ethylene and methane. However, it is best if the lab guides you on their concentrations and warning limits as these would vary for different types of oils.

4.4.4 Filter Inspection

One of the key characteristics of ESD is the polymerization of the lubricant. With this polymerization, insoluble by-products are also formed. As mentioned earlier, the lubricant typically passes through tight clearances (normally for precision applications as hydraulics or turbine oils). One of these tight clearance areas is present on the filters. Usually, filters have very small clearances, and these easily get blocked with the insoluble by-products. As such, one method of inspection can include that of the filters.

In ESD, there is also a spark involved; therefore on the filters, one can expect to see small areas of burnt filter membrane in addition to the clogged membrane. Usually, a change in differential pressure gives an indication that the membrane is being clogged.

The following can be used as a summary for the tests used to identify the presence of electrostatic spark discharge:

FTIR and QSA: Both of these tests can help to identify the presence of varnish, sludge or insoluble materials which are produced as by-products of ESD.

RULER: Typically, there is a depletion of antioxidants during this mechanism. A warning limit of <25% of the new oil value is generally observed.

DGA: With the temperatures and sparks being produced in this form of degradation, we often find the presence of acetylene, ethylene and methane with this test.

Filter inspection: During ESD, some of the by-products tend to be filtered out due to the size of the filter membrane, and parts of the membrane can actually burn when the sparks occur. Hence, the presence of burnt and clogged membrane is indicative of ESD.

4.5 ADDITIVE DEPLETION

Throughout this compilation, we have mentioned that the lubricant has a sacrificial nature especially when it involves its additive packages. Typically, there are five tests that can be used when determining if additive depletion has occurred: FTIR, Colour, QSA, RULER and RPVOT. While these are no strangers to the reader (as they have been covered in the previous pages), the results from the tests differ for this type of degradation.

4.5.1 FTIR

As we explained earlier, the FTIR is very powerful and can detect the presence of different molecules that are present in the lubricant. It is also helpful to note the absence of these molecules during testing. For instance, when testing a gear oil if the FTIR does not show anti-wear additives or reduced concentration of these, then it is safe to assume that the anti-wear additive package is being depleted. This should not be assumed in isolation. It is helpful if a trend is monitored over a period of time to determine if the anti-wear package is indeed being depleted.

4.5.2 Colour

Particular additives have definitive colours. As such, when they are depleted, their absence can be noted by the change in colour of the lubricant. This test is not definitive of additive depletion as a change in colour could also represent contamination with another source. Additionally, some lubricant manufacturers can add dye to their products. Therefore, it would be best if the lab's guidance is sought during the analysis of colour in used oil analysis.

4.5.3 QSA

In the QSA test, the presence of insolubles is measured and tested. Therefore, if the additives become depleted and deposit, the QSA test can help us determine which additive has been deposited.

4.5.4 RULER and RPVOT

Both of these tests are focused on the presence of antioxidant additives. Thus, these tests can confirm if these are being depleted or not and their rate of depletion. As indicated earlier, these rates are heavily dependent on the type of lubricant and compared against the original lubricant values.

The following is a summary of tests that can be done for additive depletion:

FTIR: This can be used to determine the presence or absence of additives. The values must be compared to that of new oil to determine if there was in increase or decrease in their concentration.

Colour: Additives have particular colours; however, this may vary depending on the additive used and the lubricant manufacturer. In this case, one should rely on the guidelines issued by the lab to identify the presence or absence of particular additives.

QSA: The presence of additives in insoluble materials that have been deposited can be determined by this test.

RULER and RPVOT: These two tests identify the presence or absence of antioxidant additive by reporting the concentration of phenols and amines which can help us understand the rate of depletion.

4.6 CONTAMINATION

When assessing the presence of contamination, there are three main tests that can be used to determine its presence. However, the user can also physically inspect the sample and compare it to that of the original lubricant to get an approximation of the existence of contamination. Although viscosity has not been included in these tests, it should be noted that the presence of contaminants can impact on the viscosity, causing it to either increase or decrease depending on the concentration and type of contaminant. The three main tests that can be used to identify contamination are described below.

4.6.1 FTIR

The FTIR test is very powerful as we have seen throughout these pages. Since it determines the elements present, it can quickly let us know of the presence of any abnormal elements. One common element that one can find may be silicon (this can be present through dust from either the atmosphere or sealant).

However, we cannot issue standard warning limits for these contaminants as they will vary greatly depending on application and environment.

4.6.2 Colour

While the Colour test may not be as definitive as the FTIR, it can certainly let us know if there are any contaminants present. When the oil degrades, it usually darkens; however, if the oil changes to a green colour (provided that the lubricant was not green initially), then we can assume that the lubricant became contaminated with a green substance. It then narrows our search to green substances that are available onsite. Most of the times if it is green, then it may be coolant, but this is dependent on a lot of factors, and a quick visual inspection of the oil sample can identify contamination by another fluid (of a different colour).

4.6.3 Presence of Water/Fuel/Coolant

Usually, labs test for contaminants through the Karl Fischer (presence of water), fuel presence or even coolant presence (depending on the environment of the oil). As we mentioned earlier, these limits may change depending on the application and the type of oil. However, the presence of any of these substances can be considered a contaminant and in some cases a catalyst for further reactions.

The following is a summary of tests that can be done for identifying contamination:

FTIR: We can use this test to determine the presence of any abnormal elements and detect large increases in contaminants such as silicon. The condemning values vary greatly for this test depending on the application and environment of the equipment. Always compare the used sample to that of new oil to determine any trends.

Colour: This is a quick identifier in determining if the lubricant has been contaminated with another (of a varying colour). In these cases, it is important to pay attention to the other fluids that are within the same vicinity of the lubricant and note their colours.

Presence of fuel, water or contaminant: Once there is fuel, water or any foreign substance, then the lubricant has been contaminated. The values differ depending on the application and environment, but the presence indicates that the foreign material did ingress and contaminate the lubricant.

4.7 SUMMARY OF DEGRADATION MECHANISM TESTS AND RESULTS

As seen in the previous pages, all of the degradation mechanisms have specific tests and results associated with them. Table 4.1 summarizes the tests that can be used for identification of these mechanisms.

TABLE 4.1 Summary of Degradation Mechanisms and Their Results

TESTS	OXIDATION	THERMAL DEGRADATION	MICRODIESELING	ESD	ADDITIVE DEPLETION	CONTAMINATION
Viscosity	Increase 5%	Decrease 5%				
Acid number	Increase 0.3 mgKOH/g					
Colour	Rapid changes	Rapid changes			Use lab guidelines	Presence of abnormal colours
FTIR	Presence of insolubles	Presence of carbonaceous deposits	Presence of soot, tars, sludge, carbon insolubles	Presence of varnish, sludge, insolubles	Presence/absence of additives	Presence of abnormal elements (water, fuel, coolant)
MPC	25–35: Abnormal, >35: Critical					
RULER	<25% new oil			<25% new oil	Presence/absence of antioxidant additive	
RPVOT	<25% new oil				Presence/absence of antioxidant additive	
Visual inspection			Cavitation present			
QSA				Presence of varnish, sludge, insolubles	Presence of additives in insoluble materials	
DGA			Presence of acetylene, ethylene and methane			
Filter inspection			Presence of burnt and clogged membrane			

Dealing with Degradation

5

Although we have identified the various methods of lubricant degradation, we also need to understand routes by which we can reduce these processes and/ or implement the aforementioned tests in the most economical way. While lubrication practices may differ widely depending on the industry/application in which it is being used, the tests remain fairly standard and can be applied in stages as discussed in the following sections.

5.1 UNDERSTANDING YOUR EQUIPMENT

The equipment in your facility is critical in ensuring that the facility is operational and meets its production rates. While it would be nice to listen to every piece of equipment and understand what is happening inside of it, this can quickly spiral depending on the number of components that are contained in each equipment. On a mid-sized power plant, this can range to around 2,000 components! Thus, one of the best ways to manage the listing is to categorize your equipment based on their criticality.

Each facility has different areas in which production can be affected. Thus, the key element in categorizing is first understanding what criticality means to you. For example, if the steam turbine goes down in a facility, it may mean that the entire plant shuts down. Thus, this is one of the most *critical* pieces of equipment on the plant and should be monitored. On the other hand, if the compressor goes down, this can mean that one part of the process does not get completed which may not affect the production levels greatly until it is time for that process to be done. Thus, this equipment can be categorized as *semi-critical*. Lastly, if the pump fails but there is a backup pump and this does not affect the production process, then this is considered *non-critical*.

Thus, each piece of equipment can be categorized as critical, semi-critical and non-critical. Now that the equipment has been categorized, we can then take a look at the components and determining the best sample points.

These vary depending on the type of equipment and its application; however, labs typically guide operators on the most appropriate sample points as per component.

Again, we can apply the critical, semi-critical and non-critical categorizations to the components after classifying the equipment. For example, in a turbine, it may be critical to get a sample from the header to understand if the temperatures are directly impacting any degradation at that point. Since this sample is critical, it will adhere to a higher frequency sampling rate since it should be monitored closely. Conversely, it may be semi-critical to collect a sample from the primary sump to fully understand the effects of the entire system on the lubricant. This sampling frequency will be less than that of a critical sampling point. On the other hand, collecting a non-critical sample from the new drum of oil as a baseline sample could occur once a year.

Thus, understanding the criticality of the machine and its components will directly impact on the frequency and types of lab tests that are conducted. By classifying them into the three categories, we can more easily prioritize our sampling programme and budget to gain a better return on investment from this form of condition monitoring.

5.2 DETERMINING THE MOST APPLICABLE LAB TESTS

Budgets are the main determining factors when deciding which condition monitoring practice can be applied. Typically, in the case of oil analysis, some of the tests are a bit expensive, while others don't dent the budget that much. While it may be convenient to have a baseline for every piece of equipment in the facility, it may not be practicable. As such, by using the categories highlighted earlier (critical, semi-critical, non-critical), we can use these as guides for allocating our budgets and the relevant lab tests.

As indicated in one of the previous chapters, the basic tests that can be performed to help us identify if degradation is present are viscosity, presence of water/fuel, TAN/TBN, concentration of additive, metals or contaminants (through spectroscopy). Most labs offer these tests as a basic combo package for relatively low prices that would not significantly impact on the budget. As such, these basic combo packages can be done on a monthly basis. With a monthly trend, we can then monitor changes in the viscosity, TBN/TAN and other elements. Depending on the trends observed (as indicated in Chapter 3), we can then recognize whether degradation is occurring or not.

5.2.1 Understanding the Results

Before we can apply the appropriate lab tests, we need to understand the results and then decide our next steps. As indicated earlier, all of the basic tests can help us in identifying that lubricant degradation is present but how do we know which follow-up tests should be applied? We can decide on the follow-up tests and their respective frequencies by understanding the results of the basic tests that were performed.

Once we have established that degradation may be occurring, we need to evaluate the environmental conditions to help us narrow the type of degradation. This can help us in choosing the next set of tests to apply (as these may be on the expensive side). Figure 5.1 provides a flowchart of the conditions required for each of the degradation mechanisms and their final deposits.

As shown in the flowchart in Figure 5.1, one of the main indicators of change is viscosity, followed by temperature and deposits (if the oil has degraded to this extent). However, when deciding the next set of tests to perform, Figure 5.2 can give some guidance.

As indicated in Figure 5.2, the first test to aid in determining the type of degradation is FTIR. This test shows the elements present in the lubricant and can lead the investigator to narrow down the possible degradation mechanisms. If there is an increase in viscosity, then the only types of degradation can be either oxidation or contamination. However, if water is present, then we can rule out oxidation. To determine the rate of oxidation or the stage at which the oxidation has reached, we can use the RPVOT, RULER and MPC tests.

If the viscosity has decreased, then this can be due to either thermal degradation or fuel being present. We must remember that both electrostatic spark discharge and microdieseling are forms of thermal degradation with varying by-products. Before sending off to the lab, one can easily verify the next test by visually inspecting the components and the filters. If the filters appear clogged with some areas of burnt membrane, then it is likely that electrostatic spark discharge is the prevalent degradation. However, by performing the Quantitative Spectrophotometric Analysis (QSA), we can determine the by-products of the degradation. The RULER and DGA can help us identify the stage at which the electrostatic spark discharge is taking place.

On the other hand, if the components show cavitation, then it is likely that microdieseling has occurred. The QSA test can help us in identifying if high or low implosion pressure occurred during the process. This can in turn assist us in determining mechanisms to reduce the microdieseling. Similarly, with additive depletion, the RPVOT and RULER tests can help us identify whether antioxidants have been depleted or not. Additionally, the DGA and QSA tests can help us determine which additives have dropped out of the lubricant.

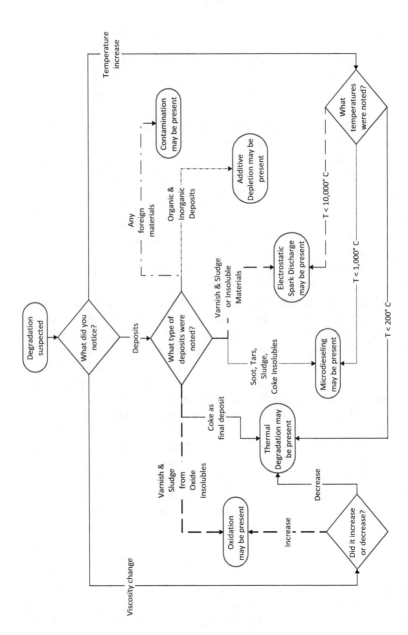

FIGURE 5.1 Flowchart of conditions required for lubricant degradation mechanisms.

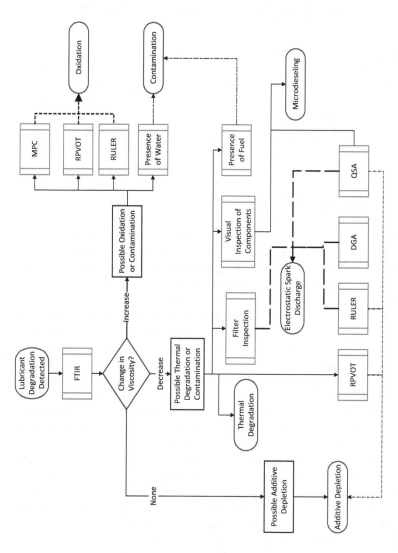

FIGURE 5.2 Flowchart to determine the follow-up tests.

5.3 IMPLEMENTING MEASURES BASED ON LUBRICANT DEGRADATION MECHANISM

Although we have covered the tests involved in identifying the degradation mechanism at work, what can we do to help alleviate the results of these mechanisms? There will be instances where our budget may not be sufficient to implement the measures or we simply cannot afford the downtime to handle the deposits or the other effects of degradation. As such, we have listed some methods of dealing with the degradation depending on the type that is being experienced.

5.3.1 Oxidation

In the initial stage of oxidation, a free radical is produced. This free radical then starts a series of chemical reactions where more free radicals are being produced. Due to the number of free radicals being produced, the antioxidants in the lubricant become sacrificial to protect it. Thus, one method of reducing oxidation would be to filter out the free radicals.

While this sounds simple, one has to understand the composition of the lubricant for your system as each system will be different. For instance, in an ammonia plant, sometimes the lubricant may come into contact with the product; however, the concentrations may differ by equipment. For most ammonia plants, there are two or three sub-plants that have identical equipment. Nevertheless, while they appear similar (and in theory, they should all operate the same), we have come across instances where the contact with the product varies. Therefore, it is critical to understand the composition of the lubricant before implementing *chemical filtration systems.*

Chemical filtration systems first analyse the lubricant in the system and identify which chemical package can help reduce the occurrence of degradation. A test sample of the lubricant in the system is reacted with the chemical package and the results are trended to provide feedback on this chemical combination. Depending on the results, the ratios can be adjusted for the system and then implemented. This is usually an ongoing process that requires an investment of time, expertise and additional equipment. The fixtures may be temporary and are designed such that it does not impact on the efficiency of the system. However, it does not solve the degradation issue; rather, it reduces its effects to allow prolonged life of the lubricant.

An alternate method to chemical filtration would be to understand the real root cause for the oxidation. Is it that the lubricant is being exposed to higher than usual temperatures, what are causing these temperatures? In most cases, the increase in temperatures occurs at the seals or servo valves where the clearances are particularly smaller. Perhaps the lubricant is not flowing as quickly as it should to reduce the build-up of heat? In these cases, we can identify if the system can be modified by *increasing clearances or adjusting lubricant flow* to reduce the temperatures. These require approvals from the OEMs and may take a longer time to implement as these are permanent changes to the system.

When investigating the root causes for oxidation, one must be aware of any changes to the system. Hence, baseline results before oxidation occurred are extremely useful in these instances. However, there are cases where baseline results may not exist and monitoring only begins when oxidation has been reported. In these cases, the rate of oxidation should be trended to establish timelines to determine expected remaining life of the lubricant or schedule downtime to clean and return the system to normal.

5.3.2 Thermal Degradation

Similar to oxidation, thermal degradation involves an increase in temperature of the lubricant. However, with thermal degradation, the lubricant actually undergoes thermal cracking where the viscosity is decreased. One method of reducing this type of degradation is to identify the source of the increase in temperature. There are instances when the temperature increases due to *changes in the process* such as a new catalyst being used or *changes in the infrastructure*, for example, new piping installed (of a different material from the previous pipes). *Any changes to the system and process* should be investigated thoroughly before implementing permanent physical changes as stated earlier.

Another method of reducing the heat build-up would be to investigate if the *flow of lubricant* is sufficient to allow cooling can the *residence time be increased* to allow more time to cool or *can clearances be increased to increase flow*? These are all questions that can be asked to each system as they will be different and operate in varying conditions, for instance, what if the operating hours were increased recently from 8 to 16 hours per day (with a double shift or even 24 hours). Did we consider what effects this would have on the lubricant? Was the system designed for these prolonged hours of operation or are there adjustments to be made when running for longer hours? By investigating the source of temperature increase, we can ideally address the reason for thermal degradation before implementing changes to the system.

5.3.3 Microdieseling

The major reason behind microdieseling is the presence of entrained air. If you couple entrained air with different pressure zones, then we have localized temperatures in excess of 1,000°C as the entrained air moves. Therefore, we need to eliminate the instances of entrained air and varying pressure zones.

When eliminating entrained air, the first aspect that we can consider is the actual formulation of the lubricant. With most lubricants, they state the *air release values, foaming tendencies* and even *emulsion times*. These are important to note when choosing a lubricant that can be subjected to microdieseling. Thus, when evaluating lubricants to be used in your system, special attention should be paid to these values. If microdieseling occurred before, note the values of this lubricant and compare them to the potential lubricants.

Another method would be to explore your system and understand if there are instances where the lubricant is being *churned too quickly* with a *short residence time*. If a lubricant does not have a longer residence time, then the air remains entrained as the lubricant passes from one pressure zone to another. It may be more difficult to change the pressure zones that the lubricant has to pass through as each component will have a different pressure. In these cases, the *residence time can be increased* to allow more timely air release. Operators can explore these mechanisms by idling the equipment or decreasing the flow rate to establish the reasons for microdieseling before making changes to the system which will require OEM approvals.

5.3.4 Electrostatic Spark Discharge

The presence of static electricity is the main component behind electrostatic spark discharge. This static may accumulate, while the lubricant is passing through the tight clearances throughout the system. One method of reducing the static is through the use of *antistatic filters*. These filters remove static as the oil passes through them, thus ensuring that there is no build-up of static in the lubricant to lead to ESD.

In some systems, it may be difficult to install these types of filters as they may not meet the required size and may change the flow rate within the system. In these instances, a *kidney loop filtration system with these antistatic filters* can be implemented. This allows for a semi-permanent fixture to be attached to the system which does not impact directly on its efficiency. These should be placed strategically in the system such that they remove the static before the oil reaches critical components. Wire mesh screens can also be placed throughout the system to help remove some of the static.

5.3.5 Additive Depletion

When additives get depleted, they are being forced out of the lubricant that they swore to protect. We therefore need to identify the reasons that they are being driven out. There are a couple of reasons for this type of depletion such as contamination or resources being used to slow down the rate of degradation from one of the other mechanisms. It is critical to identify the root cause for the depletion since only then can we implement measures to curb its effects.

For instance, if the antioxidants are being depleted rapidly, we need to understand if this is due to oxidation or electrostatic spark discharge. Conversely, if the TBN values are decreasing rapidly, then we need to investigate if there is a contaminant that is responsible its reduction. Most of the times, if fuel gets into the lubricant, it quickly decreases the viscosity and reduces the TBN levels. Thus, it is important to identify the reasons behind the depletion before implementing any measures.

Ideally, it is not advisable to replenish the additive that has been depleted by adding it (as a standalone) to the lubricant. When lubricants are formulated, they are blended to specific ratios to ensure that the finished product is balanced such that the anti-wear additive does not compete with the rust and oxidation additive as an example. By replenishing the depleted additive to the lubricant, this can cause the lubricant to become unbalanced and may cause other additives to drop out or become polymerized.

5.3.6 Contamination

This type of degradation is one of the most common types; however, contamination can be very different depending on what is contaminating the lubricant. In an ammonia plant, the contaminant could be the catalyst, whereas in a diesel engine, the contaminant could be fuel or coolant. As stated earlier, this is where *determining the source and type of contamination* is absolutely critical.

Should the source of the contamination be fuel, then there may be a leaking fuel injector or rings inside of the engine. This needs to be fixed before the source of ingress can be eliminated. However, if the contaminant is silicon, then we need to determine where this originates. For instance, if we are thinking about an engine, did we recently add some sealant where the lubricant may pass? Perhaps the silicon is being transferred into the engine when it is topped up either through dirty decanters or leaving the packaging open to the atmosphere for prolonged periods. The first step in dealing with contamination is *identification of the contaminant and its ingress.*

In most instances, contaminants enter the lubricant while it is still in storage or during its dispensing. It is therefore critical to ensure that lubricants are stored properly, away from contaminants such as air pollutants or water ingression which can occur if they are stored in dusty environment or outdoors. *Desiccant breathers can be used* to reduce the ingress of water in the machines. Proper *filtration* and *clean decanting equipment* can also significantly reduce the source of ingress for contaminants into the equipment.

Summary

6

Throughout this book, we have identified the various lubricant degradation mechanisms and looked at ways to differentiate them through their environmental conditions and deposits. We have also covered the lab tests that can be useful in their identification. These lab tests usually act as the initial stages of the equipment's conversation with the operators. The results of these tests can inform us of what the equipment has seen and been through up to that point in time of the sample being taken. It is then up to us to listen to the equipment and try to reduce these degradation mechanisms.

We have also included the secondary lab tests that can be used to properly identify the type of degradation that is occurring. These tests are usually a bit more expensive than the basic tests and are not done as frequently as the results take a longer time to be processed. With the results of these tests, we are now armed with more information to make adjustments to our maintenance schedule and make efforts to reduce the onset of any lubricant degradation.

Lastly, we looked at the methods of actually dealing with contamination. We explored the various processes that can be implemented for each of the degradation mechanisms and why these methods can assist in reducing the effects of degradation. One of the most critical points to remember when dealing with degradation is to find out the real root cause of the degradation. Unless this is known, we may not be implementing the most effective measure for the degradation that the equipment is undergoing.

Case Studies

7

<hr>

7.1 CASE STUDY 01: EXTENDING THE LUBRICANT SERVICE LIFE IN GAS TURBINES FOR MANUFACTURING APPLICATION

<hr>

Author: Sanya Mathura of Strategic Reliability Solutions Ltd

Sector: Gas turbines in a facility producing liquefied natural gas

Problem statement: Client wants to extend the life of the oil in service by a further six months past the routine change date. This will allow them to complete their required level of productions to fulfil their customer's orders. The extension of the oil life is carded to take place on their critical pieces of equipment for Train B (a series of 6 GE Frame 5 Gas turbines).

Background: Facility produces liquefied natural gas and has four trains each consisting of 6 or 7 GE Frame 5 turbines. Each train was commissioned separately and are treated as separate units with differing dates for end of service life. After seven years, each train is recommissioned with new turbines and a full overhaul completed on the entire train. For Train B, the time was approaching for the mandatory overhaul; however, the company needed the extra capacity to fulfil customer orders. As such, it decided to extend the service life of the equipment on the train but needed to ensure that the lubricant could withstand the additional time and perform successfully.

Action plan: Baseline samples were taken from each of the turbines (critical units) on the train to establish the current condition of the lubricant in service. The following tests were performed: Membrane Patch Calorimetry (MPC), Rotating Pressure Vessel Oxidation Test (RPVOT), Remaining Useful Life Evaluation Routine (RULER), viscosity, Total Acid Number (TAN), tests for contaminants (including ISO 4406), additive packages and wear metals. It was found that the wear metals were considerably low which indicated that there was minimal wear and that the lubricant was still protecting the elements. Viscosity and TAN levels were well within the acceptable limits as well as the additive packages. The contaminants were minimal, and the MPC levels were below 15. However, the RPVOT and RULER tests indicated that the oil was nearing the end of it life showing sharp decreases within the past year from an average of 72% to 47% for most of the units.

With turbine oils, the warning flags are raised when the RPVOT or RULER levels get to 25% of the original value. From trending the RPVOT and RULER values in turbine oils, it has been noticed that both the RPVOT and RULER values are initially high and then nosedive towards the 25%–35% region and taper off to remain in this region sometimes for years. As such, this drastic decrease was seen as normal for the length of time that the oil had been in service (approximately six years). However, given the risk involved of extending the life and the assurance needed for no unplanned shutdowns, the team wanted to ensure higher RPVOT and RULER values.

To increase the RPVOT and RULER values, the oil could be sweetened; however, the ratio of the new oil would have to be determined. For the sweetening process, one must ensure that the new oil is very similar to the oil in service. As such, one should verify if there were any formulation changes from the batch that was used to initially fill the equipment compared to the tentative new fill. Additionally, it should also be noted that even though the oil may be sweetened, this does not indicate that the RPVOT or RULER levels will return to that of new oil.

Given that the plant could not shut down, the sweetening of the oil had to be performed while the plant was operational. In this case, it was decided to sweeten the oil by 30%. To perform this task while the equipment is still running would indicate that an average of 30% of the sump must be drained. Before this was done, the Original Equipment Manufacturer (OEM) was consulted to determine the minimum sump levels that can be used without causing harm to the equipment while in operation. Thankfully, for this equipment the minimum levels were >40%; thus, a sump capacity of 70% during the sweetening process would not harm the equipment.

Once the oil had been removed, the new charge of oil was pumped into the sump. Ideally, the same pumping equipment should not be used as this will transfer any contaminants from the old oil into the new oil. Once the new charge was filled, we allowed the equipment to normalize and continue working for at least 24 hours before a baseline sample was taken to determine the new RPVOT and RULER values. Both values rose by approximately 20% (the new values were in the range of 68%–75%) for the units which were sweetened. These values were monitored monthly to ensure that any drastic changes could be found in time and not allow for an unplanned shutdown.

7.2 CASE STUDY 02: FUEL INGRESS IN MARINE APPLICATION

Author: Sanya Mathura of Strategic Reliability Solutions Ltd

Sector: Fleet of Marine tug boats responsible for transportation of cargo

Problem statement: The levels of fuel dilution appear to be of concern for a targeted number of vessels already on the oil analysis program.

Background: A fleet of tug boats responsible for carrying cargo off the coast of Trinidad start noticing high levels of fuel being reported on their oil analysis data. In some cases, the fuel levels reached alarming levels, while in others, the levels remained at acceptable but increased. Evidence of wear metals and other contaminants was found through these tests. The fleet consisted of mixed engines mainly Yanmar 12LAK-STE2, YX351-1, 6AYW-ETE and Caterpillar 3512s using a mineral SAE40 oil were sampled every 500 hours just before performing a lubricant change.

Action plan: The root cause for the ingress of the fuel needed to be ascertained. Sampling frequency was increased to include samples from the drum of oil, baseline sample (after 50 hours) and samples for every consecutive 50-hour interval. Viscosity, TBN, wear metals and fuel dilution were all trended during these intervals. Any instance of a top-up was recorded and a sample taken to determine if this was the root of fuel ingress.

By monitoring various vessels, it was noted that the only particular engines (12LAK-STE2) experienced high increases in fuel dilution in very short intervals, whereas others noticed minimal increases over the same interval. It was also concluded that the source of fuel ingress did not stem from the new oil or the holding tanks (as these were both tested). For these engines in particular, the viscosities of the oil did decrease but remain within acceptable range as did the TBNs. Thus, the decrease in viscosity or TBN was not flagged in these instances. Interestingly enough, only one vessel with the 12LAK-STE2 engines showed increasing levels of wear metals compared to the other vessel with the same engine (which demonstrated the increasing fuel dilution levels but no wear metals).

The vessel with the increasing wear metals was given priority and scheduled for an engine rebuild with a change of the fuel injectors. Even though the engine operated at a reduced capacity, it was still functional; however, the decision was made to have the engine repaired to avoid unplanned downtime in the future. After the rebuild with new fuel injectors, the fuel ingress levels returned to minimal for this vessel. As such, the decision was taken to replace the fuel injectors on the similar vessel to determine if they shared the same cause. After these changes, it was noted that the levels for both vessels were now minimal and the wear metals regulated.

It must be noted that when investigating the source of ingress for any contaminant, all points of entry must be examined and thoroughly checked before they are ruled out. In this instance, the oil from the drum, holding tank (new oil), sump of engines and after top-up were all checked to determine the presence of fuel. The dispensing practices should always be noted and investigated as a possible source of ingress.

7.3 CASE STUDY 03: WATER INGRESS IN GAS TURBINES FOR MANUFACTURING APPLICATION

Author: Sanya Mathura of Strategic Reliability Solutions Ltd

Sector: Gas turbines in a facility producing liquefied natural gas

Problem statement: Dangerously high levels of water ingress noticed in one particular train (series of 6 GE Frame 5 Gas turbines).

Background: Facility produces liquefied natural gas and has four trains each consisting of 6 or 7 GE Frame 5 turbines. Each train was commissioned separately and are treated as separate units with differing dates for end of service life. For Train A, during a routinely scheduled oil change, the baseline results (after the new oil was filled in the sump) were determined to contain dangerously high levels of water (in excess of 2,274 ppm in one instance) for all of the individual sumps. Since this was a new charge of oil of 10,000 gallons, the oil could not be discarded and replaced easily.

Action plan: A vacuum dehydration unit will be used to remove the excess levels of water. Since the entire train was down for maintenance, this was an opportune time to install the dehydration units in line with the sumps and have the oils circulated through the system to remove the water. Initially, samples will be taken every 12 hours to monitor the levels of the water and every 24 hours to determine if any of the additives are being depleted and the presence of any wear metals or other contaminants.

After the first 12 hours of vacuum dehydration, one of the sumps experienced a 29% decrease of water (from 2,274 to 1,622 ppm). Then, after 24 hours, this sump experienced a total of 94% decrease on water content (from the original value of 2,274 to 138 ppm) to a safe level. The dehydration system was removed from the sump after achieving the safe water content level and the water levels monitored for the next month. It was noticed that the levels did not return to the initial high levels that were first observed.

Upon further investigation, it was noted that the oil was stored in properly sealed flexibags but that the pumps used to siphon the oil into the sump were previously used for pumping water during the previous week. It was also noted that the hatch covers for the sumps remained open for cleaning work that was done on the inside of the sump in preparation for the new charge of oil. The hatches remained open for extended periods, and the sump was not inspected before the new charge of oil was filled.

7.4 CASE STUDY 04: RAPIDLY INCREASING OXIDATION IN A COMBINED CYCLE POWER PLANT

Author: Sanya Mathura of Strategic Reliability Solutions Ltd

Sector: Steam and gas turbines in power plant

Problem statement: Client is experiencing high levels of oxidation on the header return line in the power plant.

Background: Facility is a combined cycle power plant with four gas turbines and one steam turbine (all GE Frame 7s) responsible for generating power to be supplied to the grid. They have noticed that the oil is turning dark very quickly, and through their onsite lab testing facilities, they have seen evidence of high sediment build up. This sediment build up has caused the differential pressures to rapidly increase, and the filters have been changed more frequently than in the past.

Action plan: Temperatures for the various elements in the system including the storage tank, header, gas and steam turbines and bearing temperatures were recorded to note any discrepancies. It was noted that the temperatures varied from 69°C (at the header) to 125°C (at the bearing drain from the gas turbine). RPVOT, QSA, viscosity and TAN tests were done on the oil in service. The most alarming result was the QSA which stated the varnish potential at 100 while the phenol levels were non-existent. The amine levels were still at 35%, but the RPVOT value had dropped to 15%!

The oil needed to be changed and quickly to avoid any damage on the inside of the system. Luckily, the wear particles did not show any alarming rates, but the low values did start to show an upward trend over the course of one month.

Since the system could not be shut down and was due for an overhaul with two new turbines being commissioned later in the year, the plant needed to stay operational until those were commissioned.

After performing a quick inspection of the lines which experienced the highest temperatures, it was realized that there were gas leaks that vented on the lines causing rapidly increasing temperatures of the oil. However, the gas leaks could not be stopped until the plant was shut down. Lead barriers were placed between the areas of the gas leaks and the lube lines to shield the lines from the drastic increases in temperature as a temporary solution until the plant could be shut down.

A running sweetening process was done for the oil that remained in the system with an initial sweeten ratio of 40%. After this new oil was placed into the system, a kidney loop filtration system was installed to help clean up the oil that now occupied the system. Weekly checks utilizing the onsite lab ensured that the ISO ratings decreased, while a monthly sample was sent for RPVOT and QSA in an external lab.

It was found that after the temporary temperature blockages were installed that the overall temperatures of the system decreased. Coupled with the partial new charge of oil and the kidney loop filtration system, the oil was saved from being further degraded at such an accelerated rate. Additionally, this allowed the plant to continue operation and meet it required output levels until the new turbines were commissioned.

7.5 CASE STUDY 05: SPORADIC INCREASES IN VIBRATION AND TEMPERATURE LEVELS IN AN AMMONIA COMPRESSOR

Author: Sanya Mathura of Strategic Reliability Solutions Ltd

Sector: Ammonia manufacturing plant

Problem statement: Client is experiencing high levels of vibration and temperature on one of the compressors in the Ammonia plant.

Background: Facility is an ammonia plant that has been segregated into smaller plants to optimize productivity and specialize in various ammonia products.

The main sub-plants are almost identical in design; however, only one of the plants is experiencing an issue regarding vibration and temperature increases. It was noted that when the temperature spikes, the vibration levels also increase, then the temperature drops as well as the vibration levels. The oil analysis for this component showed high MPC values and decreasing RPVOT values.

Action plan: Temperatures and vibration levels are monitored continuously on this critical piece of equipment. It was noted that the shaft centreline level changed in relation to the temperatures. When the centreline moved upward, the temperatures increased. This translated to an object causing the shaft to become out of alignment for a period during which the temperatures increased. Operators noticed that there was varnish building up on the inside of the equipment. Thus, the increase in temperature would have been caused by the varnish material becoming an insulator and causing some shaft misalignment. The shaft would have continued turning and rubbing against the varnish until the varnish got dislodged. When this occurred, the shaft would resume its regular alignment and both the vibration and temperature levels would have normalized.

While understanding the concept helped to visualize what was taking place in the system, the source of the varnish needed to be addressed. After performing a 25% partial change out of the sump, an ISOPur system was used to filter the oil remaining in the system and remove some of the contaminants. An immediate decrease in temperature was noticed after a couple of hours of the ISOPur being in the system. However, the presence of ammonia in the system oil remained at dangerously high levels. The source of the ingress of the ammonia needed to be addressed.

The seal oil traps were inspected, and it was noted that the one of the seal traps was malfunctioning causing ammonia from the process to leak into the oil and contaminate it. The seals of the compressor were also found to be faulty, and these were also allowing ammonia to get into the system. The reaction of the excess ammonia in the system caused accelerated formation of varnish in the presence of regular system temperatures of 100°C. There was a planned upcoming outage during which all the issues regarding the seals would be addressed.

Continued monitoring of the MPC, RPVOT, viscosity and TAN levels were carried out on a monthly basis for the first three months to ensure that the levels of ammonia and precursors to varnish were relieved. After the first six months with minimal changes, the testing frequency was relaxed to quarterly for the RPVOT, MPC and QSA tests, while the regular viscosity, TAN and elementals remained monthly tests.

7.6 CASE STUDY 06: LUBRICATION CONDITION MONITORING CASE STUDY – AVOIDING CRITICAL ASSET DAMAGE

Author: Andy Gailey of UPTIME Consultant Ltd

Context: Food manufacturing plant producing a wide variety of snack foods.

Critical asset: Buhler Twin Screw Extruder, producing snack food pellets for drying and frying.

Lubricated component: Bespoke manufactured KissLing AG Reduction Gearbox.

The main issues with this particular equipment were

- Legacy Equipment that was approaching 12 years old
- Little history was recorded of issues in the CMMS
- Complete oil changes were planned annually, but no evidence existed other than anecdotal reports
- Lubricant was specified as VG220 H1, and the wet fill stated as 40 l.
- Food-grade lubrication (H1) only was allowed to be used to reduce consumer risk.
- The duty cycle was high on this equipment, and production runs of 7–14 days not uncommon.
- The Gearbox manufactured by KissLing AG was made to order, so the lead-time would be in the order of months; there were no spare units, although it does have a "sister" line next to it.

We didn't know enough about this critical asset to feel confident about its ongoing health especially with very little history in the Computerized Maintenance Management System (CMMS) about the lubrication management. This was identified as one of ten units on site with high criticality that I decided to carry out monthly lubrication sampling with the help of an independent laboratory, and I also reviewed the VG220 specification for the environment we were operating in. This proved to be correct; after a discussion with our lubrication supplier, I decided to retain this specification and monitor the monthly lab results.

It was fortunate that I knew this equipment since it was commissioned and the operators who looked after it, and I made sure to talk to them about any issues they could remember, and none recalled any major issues with the gearboxes. The main aspects that can degrade oil and greases in industry are excessive heat, water ingress, heavy production demand, people, incorrect specification, airborne debris, poor housekeeping practices and poor lubrication methods; the list is extensive and applies more or less in each sector. Water ingress is a larger risk in the food industry, as it was these assets were in quite a dry area although humidity could be an issue in the summer months.

I introduced monthly oil sampling and made sure I understood how to take "good" samples; I introduced an in-line sample port, and this meant I could take running samples when the extruder was in full production; this sampling point was located on a feed section that was in disturbed oil flow and prior to the filter element; this gives the best sampling for just one port. The oil condition reports indicated the major elements that could indicate wear, e.g. Fe (Iron4), Au (Aluminium), Cu (Copper), Ni (Nickel) and Pb (Lead), were the main ones I was interested in along with three more readings: H_2O (Water), V40 (Viscosity at 40°C) and PQI (particles/million). It is important to understand what these elements relate to, what deviation away from mean is acceptable and what level of particles is okay. These questions can be answered through training.

When I set up the oil sampling regime, I sent baseline samples of unused oil samples from the same batch we were using from our supplier; this is important as the laboratory needs to know what you are using and have this base sample as a reference. Our major concerns were about the internal health of the gear train not just the state of the lubricant. I always refer to this in training as the machine blood sample, and it's very similar to us giving a blood sample to ascertain our underlying health. Just like any good doctor, I researched the internals of the gearbox to understand the makeup, this gave very important insights. For instance, there were plain bearings inside made of yellow metal that included copper and other elements I could track on the sampling results.

After six months of sampling, I now had a good "signature" of the steady state of the oil in good running condition, and I learnt that the water content was zero so that reassured us that our standard breathers were good enough; if I had seen any water content, I would have introduced desiccant breathers, but they weren't required. Particles per million were in single figures; one thing I learnt from the laboratory is that they only count below 8 micron so you still have to be aware of larger particles if you think that is a risk. I would draw the sample by purging a small volume of oil and then capture the sample in a clear plastic sample bottle; these came sealed from the lab along with single use plastic tube and sealing bag to return without contamination.

After taking the sample, I would always visually inspect the oil for colour, odour, particulates that I could see, aeration that looked unusual, and anything that didn't look the same as previous samples. I did once see increased aeration through the sight glass; this is a very basic but good way of inspecting oil health daily. I sent an ad hoc sample and gave the lab a call as they were sent by post and reported back on-line; they expedited the sample and found a slightly elevated silicone content that we traced to a cover seal being disturbed. Elevated silicone equates excess aeration.

The expenditure on sampling is very low considering the value of knowing what is going on with the internals of the gearboxes while also understanding the state of the oil, this allowed us the confidence to extend the expensive oil changes to above two years. We now only carried out oil changes "on condition" not to calendar. Food-grade oils are more expensive as they have extra processing involved and in most cases have to be produced in separate facilities to industrial oils; generally, they can be ten times the cost of most straight industrial oils.

The biggest benefit of this lubrication strategy came after about four years; one of the samples showed an excessive Fe (Iron) reading. I immediately took and sent a secondary sample, and informed the lab who had reported "elevated Fe noticed" in their report. I also used a complementary condition monitoring technique that picked up a small signal of degradation on a bearing in the large step down transfer box; this was a separate unit but importantly used the same lubricant as the main gearbox to wash the large helical gear with an integrated oil spray bar. This gave us the evidence to instruct a planned outage to strip and renew the damaged bearing, and the elevated Fe was due to spalling of the inner race of this bearing.

If this bearing had carried on towards collapse, the main gearbox would certainly been at risk of major collateral damage from the shared lubrication, and the ramifications would have been huge. Lubrication in control allows you to understand your assets, buys you time if things start to go wrong, protect your supply chain and reduce expensive planned oil changes, boosts sustainability and protects this valuable and finite resource that we all depend upon.

7.7 CASE STUDY 07: ALLISON TRANSMISSION ON SCHOOL BUSES IN IOWA

Author: Michael Holloway MLE, CLS, LLA (I, II), MLT (I, II), MLA (I, II, III), OMA 1, CRL of 5th Order Industry LLC

Product: STRATA XL 15W-40 & SYSTEM PURGE

Problem areas: A school district in Iowa experienced issues with their school bus transmissions. These were susceptible to several costly problems:

1. Gear wear due to the formation of acids and deposits.
2. Sludge and varnish form due to oil oxidation, make the transmission run hotter and require more fuel to operate.
3. Excessive deposits result in loss of power, broken gears and shortened oil life.

Proposed solution: LubeMaster SYSTEM PURGE will remove deposits on the transmission components, which is directly related to a projected reduction in premature transmission failure. LubeMaster SYSTEM PURGE was used to achieve the maximum performance level of STRATA XL 15W-40 oil which meets the Allison Transmission C-4 specification. STRATA XL 15W-40 will reduce the internal friction of the gears while maintaining a boundary layer to protect the components from rust, corrosion, shock load and wear. The increase lubricating performance will lower operating temperatures and extended oil and transmission life.

Study layout: Four buses were driven for a minimum of 45 minutes. Temperature measurements were taken and averaged using a Raytek RAYNGER ST20 Pro Noncontact IR Thermometer which has an accuracy of ±2°F up to 400°F. Temperature readings were taken on the transmission housing and the oil pan. The LubeMaster SYSTEM PURGE was added to the existing oil and allowed to circulate, while the buses were driven for an additional 45 minutes. The existing oil and LubeMaster SYSTEM PURGE was drained, while the transmission was hot and STRATA XL 15W-40 was added. The buses were driven for an additional 45 minutes, and temperature measurements were taken.

Data interpretation: LubeMaster SYSTEM PURGE removed deposits in all four buses allowing STRATA XL 15W-40 to cool, lubricate and protect the transmission. Lower temperatures (15%–20%) mean extended oil life along with reduced transmission repair. A greater duration between transmission rebuilds should be realized.

7.8 CASE STUDY 08: MOBILE HYDRAULIC SYSTEMS ON MOBILE MIX TRUCKS IN TEXAS

Author: Michael Holloway MLE, CLS, LLA (I, II), MLT (I, II), MLA (I, II, III), OMA 1, CRL of 5th Order Industry LLC

Product: HOC ISO68 Hydraulic Oil & SYSTEM PURGE

Problem areas: A contractor in Texas had a fleet mobile mix trucks which were susceptible to several costly problems:

1. Hydraulic pressure increases due to deposit build-up from oxidized oil resulted in seal and line rupture.
2. Hydraulic pump castings break due to pressure spikes from line restriction.
3. Premature hydraulic pump wear due to increased levels of deposits.

Proposed solution: LubeMaster SYSTEM PURGE has been proven to remove deposits, neutralize acids and prepare the metal surfaces. This action allows the high concentration of performance surface active ingredients, i.e. anti-wear agents, rust and corrosion inhibitors, oxidation inhibitors, and metal deactivators in HOC ISO68 hydraulic oil to perform to its maximum capability.

Study layout: The initial temperature, pressure and engine measurements were made, while the Mack DL690 mobile mix concrete truck #602 hydraulic system was in operation. Temperature measurements were made using a Raytec RAYNGER ST20 Pro Noncontact IR Thermometer which has an accuracy of ±2°F up to 400°F. Measurements were taken and averaged on areas on the pump, filter and PTO housing. After several hours of measurements, LubeMaster SYSTEM PURGE was added and allowed to circulate, while the system was running for an additional 45 minutes. The existing oil and LubeMaster SYSTEM PURGE was drained while the system was hot and HOC ISO68 was introduced with new filters. Measurements were taken throughout the following day. The sack rate and yards poured were kept constant through the course of the study.

Data interpretation: LubeMaster SYSTEM PURGE removed deposits, and HOC ISO68 decreased the operating temperatures, extending oil life and establishing a greater duration between hydraulic pump rebuilds. Reduced engine rpm and operating pressure spikes will dramatically reduce system wear along with line and seal bursts.

7.9 CASE STUDY 09: INDUSTRIAL AIR COMPRESSORS IN A GLASS MANUFACTURING FACILITY IN TEXAS

Author: Michael Holloway MLE, CLS, LLA (I, II), MLT (I, II), MLA (I, II, III), OMA 1, CRL of 5th Order Industry LLC

Product: SYNCOM 30 & SYSTEM PURGE

Problem areas: A glass manufacturing facility in Texas realized that their reciprocating screw air compressors were susceptible to several costly problems:

1. Component wear due to the formation of acids and deposits.
2. Wear on reciprocating screws causes compression loss, reduced air throughput and eventual complete system failure.
3. Oxidation of oil creates sludge, carbon, soot and gum. These deposits increase the operating temperature of the system.
4. Excessive deposits result in loss of lubrication, compression and system wear.

Proposed solution: LubeMaster SYSTEM PURGE will remove deposits on lubricated components, which is directly related to a projected reduction in premature component failure. Lower operating temperatures also mean extended oil life. LubeMaster SYSTEM PURGE was used to achieve the maximum performance level of SYNCOM 30 compressor oil which has a very high concentration of surface-active agents such as extreme pressure agents, rust inhibitors, and shock load reducers. SYNCOM 30 will reduce the internal friction of the reciprocating screws while maintaining a boundary film in the compression channels.

Study layout: The 150 hp Ingersoll Rand air compressor SSR 1846-479 was in operation during initial temperature measurements. Temperature measurements were taken and averaged using a Raytec RAYNGER ST20 Pro Noncontact IR Thermometer which has an accuracy of ±2°F up to 400°F. Temperature readings were taken and averaged on areas of the motor housing, the flange and the screw casing. The LubeMaster SYSTEM PURGE was added to the existing oil and allowed to circulate while the compressor was running for an additional 45 minutes. The existing oil and LubeMaster SYSTEM PURGE was drained while the system was hot and SYNCOM 30 was introduced with new filters. The compressor was run for an additional day, and then, temperature measurements were taken.

Data interpretation: LubeMaster SYSTEM PURGE removed deposits allowing SYNCOM 30 to cool, lubricate and protect the reciprocating screws. Lower temperatures mean oil life will be extended along with reduced system repair or replacement. A greater duration between component repair and oil change intervals should be realized.

7.10 CASE STUDY 10: INDUSTRIAL HYDRAULIC SYSTEMS IN AN AUTOMOTIVE MANUFACTURING FACILITY IN INDIANA

Author: Michael Holloway MLE, CLS, LLA (I, II), MLT (I, II),

MLA (I, II, III), OMA 1, CRL of 5th Order Industry LLC

Product: HOC ISO46 Hydraulic Oil & SYSTEM PURGE

An Automotive manufacturing facility in Indiana uses five Van Dorn Demag 1,000 ton clamping capacity moulding presses with 60 oz. shot capacity, 6,000 psi maximum operating pressure. Each machine has 6 pumps, one 40 gpm variable volume PVQ Vickers pump, 4 Vickers fixed volume vane pumps all 38–40 gpm each and one 8 gpm Vickers vane pump. Main clamping cylinder is 60 diameter bore and tank capacity 810 gallons.

Problem: The machines have been in service without an oil change for 4 1/2 years. The hydraulic oil from Press #4 was tested and found to have severe oxidation. The oil also produced a very acrid smell which would also indicate that it was rich in oxidation by-products. Oxidation can lead to increased pump wear, high operating temperatures and unscheduled downtime. Controlling the acid formation will control oil oxidation.

Proposed solution: LubeMaster SYSTEM PURGE has been proven to remove deposits, neutralize acids and prepare the metal surfaces. This action allows the high concentration of performance in HOC ISO46 hydraulic oil to perform at maximum capability.

Study layout: The initial temperature and ampere measurements were made, while the system was in operation before and after the System Purge and HOC ISO46 application. Temperature measurements were made using a noncontact IR thermometer. Measurements were taken and averaged on areas on the pump housing, on the flange and on the casing. Ampere (colour legs) measurements were taken from the exclusive circuit box (3 phase, 460 V) by two different maintenance shifts. Oil analysis samples were taken and tested for wear metals, oil condition and acid neutralization ability (TAN – target closest to 0).

LubeMaster SYSTEM PURGE was added to the existing oil (Mobil DTE) and allowed to circulate for an additional 24 hours while the system was running. The existing oil and LubeMaster SYSTEM PURGE was drained, HOC ISO 46 was introduced with new filters.

Results: Measurements were made 20 days after the System Purge, old oil was drained and HOC ISO46 was added. There was a 9% reduction of the averaged pump and valve temperature. The analysis of the oil indicated a drastic increase in the new oil's ability to neutralize acids. The peak ampere draw, which occurs once during each cycle, was reduced 27%.

Data interpretation: The decrease in operating temperature (9%) and energy consumption (27%) indicates that the system is operating more efficiently. The low TAN values (decrease by 55%) suggest that the oil will be able to neutralize more acid and oxidation by-products over a longer period of time which will extend the life of the oil and equipment dramatically.

References

Barnes, Mark. 2003. "The Lowdown on Oil Breakdown." *Practicing Oil Analysis Magazine*, May–June, 4.

Bureau Veritas. 2008. "Basics of Oil Analysis." Brochure. Accessed May 03, 2020. http://oil-testing.com/brand-resources/pdfs/Basic-oa-2.0.pdf.

Chemistry LibreTexts. 2020. "20.3: Aldehydes, Ketones, Carboxylic Acids, and Esters." Accessed March 25, 2020. https://chem.libretexts.org/Bookshelves/General_Chemistry/Book%3A_Chemistry_(OpenSTAX)/20%3A_Organic_Chemistry/20.3%3A_Aldehydes%2C_Ketones%2C_Carboxylic_Acids%2C_and_Esters.

Chemistry LibreTexts. 2019. "Peroxide." Accessed March 25, 2020. https://chem.libretexts.org/Bookshelves/Ancillary_Materials/Reference/Organic_Chemistry_Glossary/Hydroperoxide.

Fitch, Bennett. 2015. "Identifying the Stages of Oxidation." *Machinery Lubrication Magazine*, May–June, 40.

Lewand, Lance. 2003. "Using Dissolved Gas Analysis to Detect Active Faults in Oil-Insulated Electrical Equipment." *Practicing Oil Analysis*, March–April, 62.

Livingstone, Greg and Brian Thompson. 2005. "New Varnish Test Improves Predictive Maintenance Program." *Practicing Oil Analysis*, July–August, 16.

Livingstone, Greg, Dave Wooton, and Brian Thompson. 2007. "Finding the Root Causes of Oil Degradation." *Practicing Oil Analysis*, January–February, 36.

Menezes, Pradeep L., Carlton J. Reeves, and Michael R. Lovell. 2013. *Tribology for Scientists and Engineers: From Basics to Advanced Concepts*. New York: Springer.

Noria Corporation. 2012a. "The Critical Role of Additives in Lubrication." *Machinery Lubrication Magazine*, May–June, 34.

Noria Corporation. 2012b. "Base Oil Groups Explained." *Machinery Lubrication Magazine*, September–October, 52.

Noria Corporation. 2017. Lubrication Regimes Explained. Accessed April 28, 2020. https://www.machinerylubrication.com/Read/30741/lubrication-regimes.

Scott, Bob. 2013. "Practical Oil Analysis." Paper presented at Noria Training, Double Tree Hilton, Orlando, FL, USA, September 23–27, 2013.

Sedelmeier, Greg. 2012. "Managing Water Contamination." Paper presented at Noria Reliable Plant Conference & Exhibition, Indianapolis, IN, USA, May 01–03, 2012.

Troyer, Drew. 2004. "Looking Forward to Lubricant Oxidation?" *Practicing Oil Analysis Magazine*, March–April, 46.

Wooton, Dave. 2007. "The Lubricant's Nemesis - Oxidation." *Practicing Oil Analysis Magazine*, March–April, 26.

Index

Note: **Bold** page numbers refer to tables; *italic* page numbers refer to figures and page numbers followed by "n" denote endnotes.

additive depletion 8, 10, 14–15, 22, 34, 35, 38, 41, 42, 43, 47
aldehydes 8, 10

boundary lubrication 3–5
by-products 8, 10–12, 21, 25, 26, 31–34, 41, 64–65

Calorimetric Patch Analyser 27
carboxylic acids 8, 10
catalytic fines 23
chemical filtration systems 44
colour 25, 26, 29, 30, 34, 35, 36, **38**
contamination 8, 15–16, 18, 20, 34, 35, 36, **38**, 41, *42*, *43*, 47, 49, 59
Crackle test 19

deposits 11, 12, 13, 14, 15, 30, 31, **38**, 41, *42*, 44, 49, 61, 62, 63, 64
DGA (Dissolved Gas Analysis) 32, 33, **38**, 41, *43*

elastohydrodynamic lubrication 3, 4
electrostatic spark discharge 8, 13, *14*, 32, 33, 41, *42*, *43*, 46, 47

FTIR (Fourier-Transform Infrared Spectroscopy) 25, 26, 29, 30, 31, 32, 33, 34, 35, 36, **38**, 41, *43*

hydrodynamic lubrication 3–4
hydroperoxides 8, 10

initiation *9*, 13
inorganic 15, *42*
insolubles *12*, 13, *14*, 26, 27, 29, 31, 34, **38**, *42*

ketones 8, 10
kidney loop filtration 46, 56

lacquer 9, *12*, 30
lubrication regimes 3–5

microdieseling 8, 12, 13, *14*, 15, 30–31, **38**, 41, *42*, *43*, 46
mixed lubrication 3–4
MPC (Membrane Patch Colorimetry) 25, 27–29, **38**, 41, *43*, 51–52, 57

organic 8, 15, 26, *42*
oxidation 7–12, 13, 15, 19, 21, 25–29, 30, **38**, 41, *42*, *43*, 44–45, 47, 51, 55, 61–62, 63, 64–65

polymerization 11, 13, 18, 32, 33
propagation 9

QSA (Quantitative Spectrophotometric Analysis) 31–32, 33, 34–35, **38**, 41, *43*, 55, 56, 57

radicals 9, 13, 21, 26, 32, 44
residence time 32, 45–46
resins 13, *14*, 31
RPVOT (Rotating Pressure Vessel Oxidation Test) 10, 25, 28–29, 34, 35, **38**, 41, *43*, 51–52, 55–56, 57
RULER (Remaining Useful Life Evaluation Routine) 10, 25, 28–29, 32, 33, 34, 35, **38**, 41, *43*, 51, 52

sludge 10, 12, *12*, 13, *14*, 27, 31, 32, 33, **38**, *42*, 61, 63
soot 12, *14*, 31, **38**, *42*, 63
static electricity 13, 32, 46

tars 12, *14*, 31, **38**, *42*
termination *9*

thermal degradation 8, 10–12, 14, 15, 17, 18,
      22, 29–30, **38**, 41, *42*, *43*, 45
TAN (Total Acid Number) 17, 20–21, 40,
      51–52, 55, 57, 64–65
TBN (Total Base Number) 17, 20–21, 40, 53

varnish 10, 12, *12*, 13, *14*, 27, 32–33, **38**, *42*,
      55, 57, 61
viscosity 2, 7, 10, 11, 13, *14*, 17, 18–19, 20, 21,
      22, 25, 26, 29, 30, 35, **38**, 40, 41,
      42, *43*, 45, 47, 51–52, 53, 55, 57, 59